Paumanok Publications, Inc.

January/February 2003

Passive Component Industry

An affiliate publication of the ECA
A sector of the Electronic Industries Alliance

The Only Magazine Dedicated Exclusively To The Worldwide Passive Electronic Components Industry

The Changing Market for Equipment Used to Manufacture Passive Electronic Components

INDUSTRY FIRSTS
from VISHAY INTERTECHNOLOGY

SATURATION CURRENT UP TO 118 AMPS IN A 5050 PACKAGE

Power Inductor – IHLP-5050EZ-01*

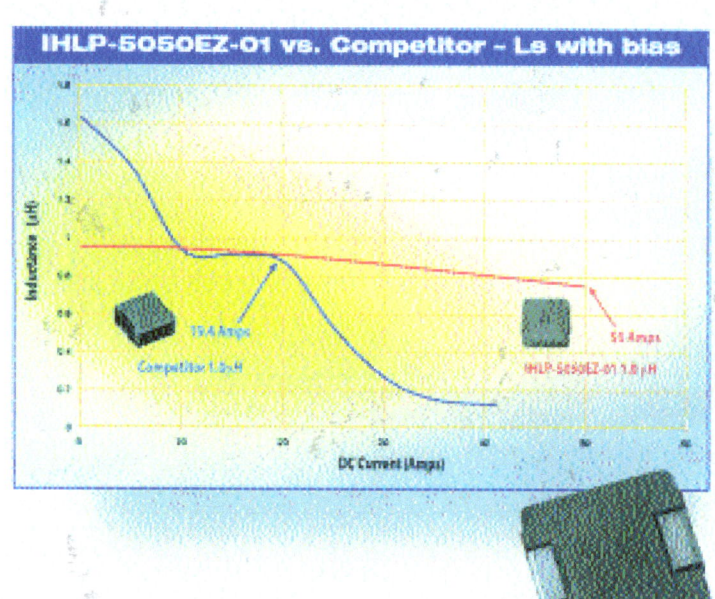

- Wide value range from 0.10 µH to 10 µH
- Handles high transient current spikes without saturation
- DCR as low as 0.53 mΩ
- Frequency range up to 5.0 MHz
- Shielded construction
- 100% lead (Pb) free
- Ultra low buzz noise due to composite construction

Datasheet at www.vishay.com/ref/IHLP

*Patented

ONE OF THE WORLD'S LARGEST MANUFACTURERS OF DISCRETE SEMICONDUCTORS AND PASSIVE COMPONENTS

DISCRETE SEMICONDUCTORS: Rectifiers • Small-Signal Diodes • Zener and Suppressor Diodes • MOSFETs • RF Transistors • Optoelectronics • ICs
PASSIVE COMPONENTS: Capacitors • Resistive Products • Magnetics
INTEGRATED MODULES: DC/DC Converters
STRESS SENSORS AND TRANSDUCERS

www.vishay.com

Your Partner in Overvoltage Protection

KEKO VARICON

Automotive Varistors

- **AV** – Leaded Automotive Varicons and Disc Varistors
  - Supply Voltage : 12 V, 24 V, 42 V
  - Operating Voltage Range : 16 to 56 Vdc
  - WLD Energy Rating: up to 10 x 100 J
  - Maximum Operating Temperature +125 °C
  - Jump Start Capability (5 min)

- **AV** – SMD Automotive Varicons
  - Supply Voltage : 12 V, 24 V, 42 V
  - Operating Voltage Range : 16 to 56 Vdc
  - 6 Chip Model Sizes : 0805, 1206, 1210, 1812, 2220 and 3225
  - WLD Energy Rating: up to 10 x 50 J
  - Jump Start Capability (5 min)
  - Maximum Operating Temperature +125 °C
  - Short Response Time

- **OV** – Dual Function Automotive Varicons
  - In both SMD and leaded Version
  - Provide Protection against Transient Surges and filter RF Interference
  - Supply Voltage : 12 V, 24 V, 42 V
  - Operating Voltage Range : 16 to 42 Vdc
  - WLD Energy Rating: 10 x 6 or 12 J
  - Jump Start Capability (5 min)
  - Capacitance Range : 470 nF to 4.7 µF
  - Capacitor Temperature Characteristics : X7R, Z5U

At least 10 %
- better performances,
- shorter delivery time,
- faster custom design,
- .....

...not the biggest, simply the best !

For more info contact:
KEKO VARICON, Slovenia
SI-8360 Žužemberk,
Phone: +386-7-3885-120
Fax: +386-7-3885-158
e-mail: info@keko-varicon.si
web site: www.keko-varicon.si

strategic partner for marketing and distribution in North America:
SEI ELECTRONICS, Raleigh, NC
Call toll free: 1-888-sei-sei-sei
e-mail: marketing@seielect.com
web site: www.seielect.com

Kamaya Carbon Composition Resistors
(High Surge @ Low Total Cost)

If you still need the high surge capabilities of this trusted resistor type, give Kamaya a call. You can depend on our consistent quality for demanding surge-handling applications for carbon composition resistors like:

- Telecom/Datacom
- Industrial motion/automation
- Medical
- Power Conditioning
- Appliances
- HVAC
- Defense
- Automotive

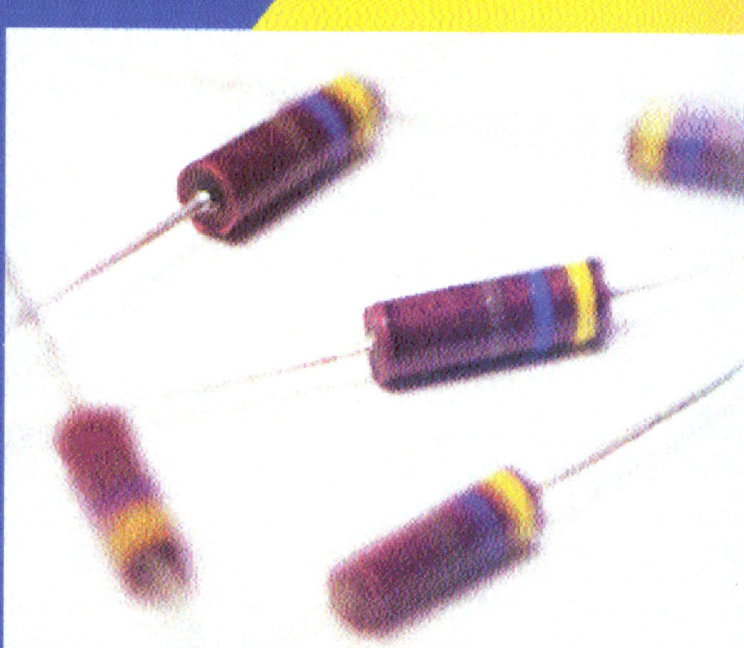

For over 50 years, we have manufactured our own resistive products to maintain the highest standards, and to be the industry's most dependable supply source. Get the best from the best: Carbon composition from Kamaya.

FREE Samples — Go To Kamaya.com For

Contact Kamaya, Inc. for technical assistance or pricing at 260-489-1533.
Fax: 260-489-2261   www.kamaya.com.

KAMAYA ELECTRIC COMPANY, LTD.

A Subsidiary of MITSUBISHI MATERIALS CORPORATION

TABLE OF CONTENTS

Passive Component Industry

JANUARY/FEBRUARY 2003   Volume 5, No. 1

The Only Magazine Dedicated Exclusively To The Worldwide Passive Electronic Components Industry

.... **FEATURE STORIES**

6   **The Changing Market for Equipment Used to Manufacture Passive Electronic Components**
The trend in passive electronic component manufacturing for the mass market is the movement away from high cost regions such as the United States, Western Europe and Japan toward lower cost production sites in China, Eastern Europe and Brazil.

10   **Advancements in Tape Casting Machine Technology**
Application diversity and new product developments are the driving forces for recent advancements in tape casting machine technology.

DEPARTMENTS ....

4   **Letter from the Publisher**
2002 Year In Review: Global Market

5   **Letter from ECA**
Electronics Flow Wheel Promises Better Industry Forecasting & Adjustment

14   **Featured Technical Paper**
Automated Solutions for Passive Component Dicing

23   **Featured Technical Paper**
Improvements in Electrode Technologies

32   **Newsmakers**
New product offerings and important developments in the passive components industry.

Cover Photo: Courtesy of Disco Corporation.

Letter From the Publisher

2002 Year In Review: Global Market

For the global passive component industry, 2002 was a year of depressed market value that continued the downward trend experienced in 2001. Continual and severe price erosion for all passive components became the norm as passive component manufacturers competed aggressively with each other in an attempt to maintain capacity utilization levels at a marginal 50%. Manufacturers offered components at prices that were, in many instances, less per part than the cost of the raw materials of which they were comprised.

The interesting aspect of this particular industry cycle is how long the market has been depressed (24 to 27 months depending upon product sector), which is the longest instance of market depression that most in the industry can remember.

Some of the trends and directions evident since 1999 continued into 2002, but at a more extreme level of urgency. These primarily related to the transference of production and market focus to Asia, with emphasis upon China, which seems to defy global economic trends and represents a beacon in the vast ocean of economic uncertainty.

As we enter into 2003, the trend with respect to the United States, Western Europe and Japan is to relocate mass market passive component production assets to low-cost manufacturing areas, and to explore the possibility of converting assets in high cost regions to value-added and application-specific passive component production.

In raw materials, last year saw the continued price erosion in the price of palladium, which reacted to the slowdown in the electronics sector, and somewhat slowed the general trend in the industry to replace precious metals with base metals for electrodes and terminations. In the latter part of 2002, palladium prices exhibited unusual price reductions (palladium is now at 25% of its price noted in January of 2001) as rumors circulated in the metals sector that Norilsk, the large nickel/palladium mine in Russia, made a bid to buy the Stillwater palladium mine in Montana, USA. The bid reportedly included cash and more than 800,000 Troy ounces of palladium in exchange for control of the mining operations.

In nickel electrode powders we saw a rather sharp increase in demand for very small particle size powders in the 0.3, 0.2 and 0.1 micron range, which are used almost exclusively in extremely high capacitance ceramic chip capacitors > 47 µF.

In tantalum metal powders the year brought tremendous infighting among competitors in the supply chain. Conflicts among capacitor manufacturers and raw material suppliers (with emphasis here on Cabot, Kemet, Vishay and AVX) resulted in public lawsuits regarding the specific nature of contracts signed in the 2000 boom year.

As we entered 2003 however, it became apparent the companies fighting in the tantalum supply chain began to realize that their public disputes, however good and righteous they seemed to be to the individual companies, were in fact bad for the overall image of the tantalum capacitor business. This resulted in an unwanted increase

Continued on page 20

PUBLISHER
DENNIS M. ZOGBI

DIRECTOR OF ADVERTISING
SAM COREY

EDITOR
JOHN D. AVANT

ART DIRECTOR
AMY DEMSKO

ADVISORY BOARD

Glyndwr Smith
Vishay Intertechnology, Inc.

Ian Clelland
ITW **Paktron**

Pat Wastal
Avnet

Jim Wilson
MRA Laboratories

Michael O'Neill
Heraeus Inc.

Daniel F. Persico, Ph.D.
KEMET Corporation

Bob Gourdeau
BCcomponents

Jack Bush
SEI Electronics, Inc.

Editorial and Advertising Office
130 Preston Executive Drive, Suite 101
Cary, North Carolina 27513
(919) 468-0384 (919) 468-0386 Fax
www.paumanokgroup.com

The Electronic Components – Assemblies – Materials – Association (ECA) represents the electronics industry sector comprised of manufacturers and suppliers of passive and active electronic components, component arrays and assemblies, and commercial and industrial electronic component materials and supplies. ECA, a sector of the Electronic Industries Alliance, provides companies with a dynamic link into a network of programs and activities offering business and technical information; market research, trends and analysis; access to industry and government leaders; standards development; technical and educational training; and more.

The Electronic Industries Alliance (EIA) is a federation of associations and sectors operating in the most competitive and innovative industry in existence. Comprised of over 2,100 members, EIA represents 80% of the $550 billion U.S. electronics industry. EIA member and sector associations represent telecommunications, consumer electronics, components, government electronics, semiconductor standards, as well as other vital areas of the U.S. electronics industry. EIA connects the industries that define the digital age.

ECA members receive a 15% advertising discount for *Passive Component Industry*. For membership information, contact ECA at (703) 907-7070 or www.ec-central.org; contact EIA at (703) 907-7500 or www.eia.org.

Letter From ECA

Electronics Flow Wheel Promises Better Industry Forecasting & Adjustment

Since anyone can remember, we've talked about an electronics supply channel, chain or pipeline – something with a definite beginning and end. This model assumes a linear procession of activities, like a relay race. It's orderly, neat and easily comprehensible. But does it reflect the product flow in the electronics components industry?

ECA believes the real model looks more like a wheel, mo varying speeds and even d directions at different times, connectivity among diffe ments. The transition bet the chain and what ECA call *electronics flow wheel* h been happening for the las decade, but it only became obvious over the last two years, when ignoring in- dustry interdependencies led to inventory and sup- ply debacles.

"The traditional con- cept of a supply channel is dead," says Glyndwr Smith of Vishay Intertechnology ECA past chair. "The su ply/demand equation can longer be addressed as a s progression. Instead there electronics flow where ea interacts with and affects every other element. Collaboration by the organizations representing different industry segments is needed to create greater transparency and cooperation within the key elements of this wheel, which will benefit everyone in the industry."

The new industry scenario provides two choices for industry sectors such as manufacturing, EMS and distribution: They can isolate themselves and participate only in their channel, chain or pipeline. Or, they can work to understand what happens throughout the electronics flow wheel, define their roles, and cooperate with the other players to make sure we don't have a replay of 2001.

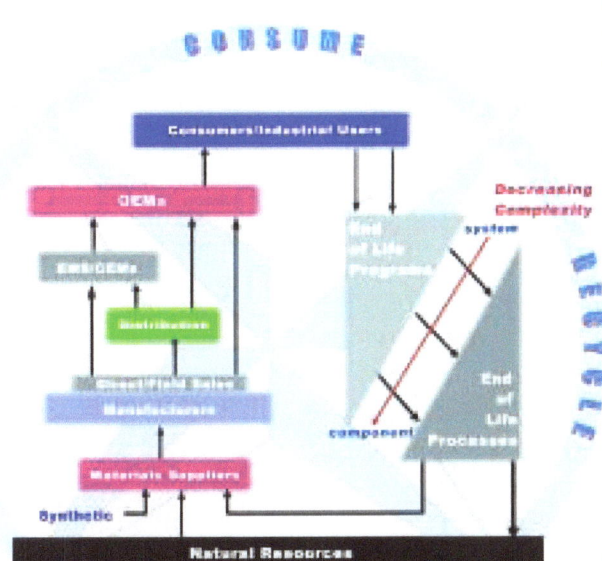

Getting Raw Materials Into the Flow

The best place to start in understanding the electronics flow wheel is raw materials. ECA has long maintained that the industry has to cognize the role of raw material ppliers. We should not be so s to believe that these will react automatically to in- ry demands. We have to look urther than two years ago ien tantalum was in short sup- ply and the flow wheel stopped and then backed up. At the same time, stories began circulating about the source of various raw materials and a possible boycott of the very natural resources needed to develop them. It is not difficult to see what kind of effect this would have on the industry.

In the linear model, the eline generally begins with introduction of raw materials the manufacturing process – manufacturing marks the fin- isn of raw materials. Yet, all of the steps that could apply going forward in the manufacturing cycle probably apply going backward in the supply of raw materials. In the traditional model, many of these steps are never taken into account, creating another gap in the flow wheel. ECA is in a position to bridge this gap and include raw materials into the flow represented in the Produce segment of the wheel.

ECA's membership already includes producers and suppliers of materials used to manufacture, assemble

Continued on page 9

FEATURE

The Changing Market for Equipment Used to Manufacture Passive Electronic Components

Abstract

The trend in passive electronic component manufacturing for the mass market is the movement away from high cost regions such as the United States, Western Europe and Japan toward lower cost production sites in China, Eastern Europe and Brazil. Consequently, existing passive component production assets located in high cost production regions are being transformed to produce value-added and application-specific passive components that have higher average unit prices and higher operating margins. Suppliers of equipment used in the manufacturing of passive electronic components should be mindful of this trend, adjust their marketing programs to accommodate this change, and change their product mix accordingly, so that they might support and profit from the changing market environment.

Current Market Environment

Following the banner year of 2000, demand for equipment used in the production of passive electronic components (i.e. kilns, tape casters, slurry mixers, anode pressers, film winders, foil winders, termination machines, etc.), declined dramatically in accordance with the decline in net new product demand for passive components in general. One prime example of this was Tokai, the Japanese producer of high fire, stabilized atmosphere kilns used in the production of base metal electrode MLCC. Tokai had a 24-month waiting period

Tantalum Powder Used in the Manufacturing of Tantalum Capacitors.
Courtesy: Sons of Gwalia

during the boom months of 2000, a lead time that literally disappeared in 2001 as ceramic capacitor manufacturers cancelled orders en masse when the market for their products declined so quickly and dramatically they could not justify any continued expansion plans. This sudden and rapid decline in net new demand for passive electronic components, which began in October of 2000 and lasted throughout 2001 and into 2002, created a classic market situation for the equipment suppliers. They helped their customers increase production capacity in 2000 to such a high level that further needs for expansion were curtailed, probably until the latter part of 2004.

In 2001 and into 2002, demand for passive component production equipment changed dramatically. The high-dollar product shipments, typically equipment that can handle large production runs (i.e. tunnel kilns, multiple tape casting lines, etc.), disappeared almost entirely, with the overall market for production equipment assuming a much lower threshold for product sales. The mix moved in favor of smaller opportunities that centered around smaller product sizes (i.e. 0201 MLCC, or the P and J size tantalum chips), or to variations of component configurations, such as the multichip array.

Movement Toward China

As prices for passive components began to decline dramatically in 2001 and price competition persisted throughout 2002, manufacturers of passive electronic components in Japan, Western Europe and the United States began to relocate the production of their standard parts for the mass commercial market to low-cost regions, with emphasis upon China, Hungary and Brazil. This movement did not entail the acquisition of new production equipment, rather it simply required

Continued on page 8

Continued from page 6

the relocation of existing capital equipment to those regions. AVX, Kemet, EPCOS each announced the development or expansion of existing assets in China for production of passive electronic components, and one of the motivating factors for Vishay's purchase of BC Components was the existence of their plants in China. For equipment manufacturers, this exodus by their customer base created a new level of competition. Industrial companies in China, Taiwan and Korea began to compete aggressively on price with passive component production equipment suppliers from the United States, Germany, Italy and Japan.

Another factor that contributed to the movement of mass produced passive components to China, Hungary, Brazil and other low cost production regions was the rapid rise of the contract electronic manufacturers (i.e. Solectron, Flextronics, SCI/Sanmina, Jaybil, Celestica, etc.), who also moved their assets to low cost manufacturing to support their small operating margins. Thus, passive component manufacturers thought it wise to locate assets in close proximity to their customer base (CEMs like to source components locally), and to also take advantage of the lower costs associated with producing in these regions.

Reorganization of Production Assets in High Cost Regions

With the movement of production assets for high volume, low priced, standard passive components to low cost regions, manufacturers of passive components began to reorganize their assets in high cost production regions. Considerable investment in brick and mortar structures had been made for years in high cost production regions when global competition in local markets was less intense because of trade barriers, tariffs and cultural ties within the supply chain. This began to change as price erosion gripped the industry. Survival of individual businesses and the importance of shareholder value (for public firms) had greater importance than historical relationships within the supply chain. The bottom line ruled.

In 2003, the trend toward production of value-added and application-specific passive components in high cost regions seems to be the rule, and equipment manufacturers should plan for and profit from this fundamental change in production. By their very nature, value-added and application-specific passive components have higher average unit prices and higher operating margins when compared to standard parts, and therefore their production in high cost regions such as Japan, the USA and Western Europe can be justified. Moreover, they represent both a challenge and an opportunity for passive component production equipment manufacturers. The higher priced components that will be produced in high cost regions will have substantially lower production volumes when compared to the mass produced parts now moving to low cost regions of the world. One opportunity for equipment manufacturers will come in offering products that are designed to produce small volumes of components, but at high yield ratios, with outstanding quality controls. But the true opportunity will come in supplying test equipment for their customers in high cost areas. Both value-added and application-specific capacitors are by nature high-reliability or specialty devices used in expensive capital goods that require quality and lifetime testing to guarantee survivability in harsh environments.

Value-added and Application-specific Passive Components

In addition to having higher prices and higher operating margins, both value-added and application-specific passive components are also more important to the design engineering and research and development operations at the passive component houses, which are assets typically located in high cost production regions. Value-added components offer standard values, but are hardened against external stimuli such as high heat (to 200 degrees C), thermal shock, vibration, overvoltage, exposure to water, humidity, oils, salts, etc. They also include standard values that operate with low ESL, and low ESR. Application-specific passive components are defined as non-standard parts with respect to voltage (i.e. >500 Vdc), or frequency (> 1 Ghz). The parts with the highest prices and greatest operating margins are those that are non-standard parts with respect to voltage or frequency that must be hardened for specific applications in the defense, medical, mining and oilfield services industries. Such components require individual part attention and testing, and require production and test equipment to meet those requirements.

Trends & Directions

Passive component equipment manufacturers will continue to see their customer base move high unit volume production assets to low cost regions to maintain their margins. Western and Japanese passive component production equipment manufacturers will continue to see Chinese and Korean competition in equipment development and sales in these regions, at prices that are lower than they can afford to produce these machines in their home countries. Opportunities still exist for Western and Japanese equipment producers in their home markets, as their customer base in those regions convert their existing brick and mortar assets to value-added and application-specific parts. These conversions require attention by the equipment manufacturers who can supply production and test equipment designed to service those sectors. ❑

Continued from page 5

and maintain electronic components and assemblies. We have increased our emphasis in this area by dedicating a board of directors position to material suppliers. We will also initiate a program that will develop information to assist material members and incorporate their data with that of other participants in the flow wheel.

Recycling or Back to Earth

It is interesting that many of the stories on the raw materials dilemma were published on recycled paper. Recycling is an issue with which our industry must come to grips. The process of design for re-use or disposal is often removed from the traditional concept of a supply chain, channel or pipeline. Most programs are focusing on system-level recycling or disposal that has more impact for the OEM end product. Very soon, however, electronic components will need to be manufactured for recycling. This again points to a more circular product evolution than a linear one.

Through its affiliation with the Electronic Industries Alliance (EIA), ECA addresses environmental issues associated with product disposal and recycling. These issues will no doubt begin to assume greater urgency for manufacturers as they migrate deeper into the product development process.

Creating Flow Transparency

To place a twist on the real estate credo, there are three things that are important to electronics industry forecasting: communication, communication, communication. At the ECA Spring 2002 meeting, participants in "The Great Inventory Train Wreck of 2001" session were almost unanimous in their belief that communication among the participants in the supply cycle could have tempered the steep decline experienced by the industry. If only each party had provided information to the others, then most felt that moderation would have entered the equation and the mass oversupply could have been averted.

Further insight into this situation indicated that much of the information being provided was not accurate or had been misinterpreted by the participants. What was truly needed was the ability to see and interpret the information as a group. Then, each party could make the best decisions for their interests and the overall prosperity of the industry.

The problem isn't the amount of information available, but providing it in an easily usable form. Organizations exist that collect, filter and disseminate information, but more cooperation among them is needed. ECA is working to make this happen on an ongoing, formal basis in 2003.

Cooperation among existing organizations takes us a long way toward better understanding information and putting it in context, but there is still a missing piece. No organization yet exists to represent EMS or CEMs. ECA is moving to correct this situation by hosting a forum to address the issues within the EMS industry and provide a platform to organize participants. If this can be accomplished, the next steps are to add to the information that exists, analyze and filter the data, and provide a new level of transparency into the electronics flow wheel.

Making Collaboration a Priority

In the linear model, organizations and participants are responsible for their segments of a channel, chain or pipeline. In standalone situations,

Continued on page 31

FEATURE

Advancements in Tape Casting Machine Technology

*James W. Dennis, Pro-Cast Division,
HED International, Inc., Ringoes, NJ*

The last century has seen the development of tape casting technology and the fabrication of thin layers of materials to produce single layer or stacked and laminated multilayer structures. Today, tape casting is the basis of the fabrication process that is routinely used in the production of multilayer ceramic capacitors (MLCC), low temperature co-fired ceramics (LTCC), lithium batteries, fuel cells and many new microelectronic devices.

The diversity of applications and new product developments has become the driving force for many of the recent advancements in tape casting machine technology. Tape casting machines have become more automated. Other modifications have improved quality and made fabrication easier. The latest tape casting machines, (Fig.1) produce better tape with more uniform thickness and much thinner dimensions. Tape cast parts now

Figure 1

can be made as thin as 1 to 2 microns and as thick as 6 mm. These can be produced to widths that exceed 2 meters and lengths that are only defined by production schedules. This versatility facilitates subsequent procedures to fabricate the final shape or part, many of which could not be produced by any other processing technique.

This new generation of tape casting machines integrates a sturdy structure with a solid, level platform that uniformly supports a moving casting surface that may consist of a steel belt, polymer carrier or both. Totally enclosed machines provide a clean room environment to minimize undesirable air infiltration and contaminants. Viewing doors with tempered glass windows and seal-tight gaskets allow fast, easy access to the casting station and all other functional areas. In addition, the machines are designed to deliver conditioned casting material consistently to the casting/coating station. The main components of the casting station are the casting head, the casting surface and the web, carrier or steel belt. The casting head can be a slot die, roll coater, adjustable doctor blade or comma bar that is precisely made to allow work to tolerances of less than 1 micron. The casting surface substrate can also be a high precision plate such as granite, certified flat to 50 millionths.

To produce a steady stream of uniform tape requires a stable flow of casting material, as well as perfect control of the casting speed and drying conditions. Precise control of the tension and speed of the moving carrier or steel belt conveying the tape through the entire tape casting machine is essential to ensure uniformity. This is accomplished by using tension monitors and a digital tachometer with a microprocessor-controlled drive and closed-loop feedback system.

The importance of properly drying the cast tape is a major concern and factor in machine design. The result is that the drying oven on new tape casting machines is carefully designed to maintain a stable temperature profile, atmosphere saturation gradient and air flow. Drying is done in two ways. Heat is gradually applied below the tape while a carefully controlled flow of heated atmosphere is passed over the tape surface. It has been found that more perfect dry tape is produced when these oven conditions are precisely controlled and can be tailored for any tape thickness and drying characteristic. The objective is to dry the tape as thoroughly and as quickly as possible while preventing uneven drying, such as the formation of a dry skin on the exposed surface of the tape. This defeats sub-surface drying and

Continued on page 12

If Variety Is The Spice Of Life...
Then Let NIC Add Flavor To Your Design!

NIC offers over 100 different product series of passive components! With so many choices, NIC gives you the options you need to spice up your designs...

If you need lower ESR, reduced package sizes, high frequency performance or extended temperature ratings, NIC has a series to meet your needs!

With so many component options to choose, NIC can show you the value and ease of reducing or consolidating your multi-supplier bill of materials.

With over 5 billion surface mount and leaded components in our global inventory we can support your expanded production requirements.

Check out NIC's full feature website:
www.niccomp.com

- *For the latest product updates*
- *On-line cross referencing*
- *On-line stock checks (distributor inventory)*
- *On-line sample requests*
- *Product selection guides*
- *Technical product info*
- *Quick look-up guides & tools*

Corporate Headquarters
70 Maxess Road
Melville, New York 11747
(631) 396-7500 / FAX 631-396-7575

Western Region
2070 Ringwood Avenue
San Jose, California 95131
(408) 954-8470 / FAX 408-954-0349

NIC Components Corp.

Technical Inquires: tpmg@niccomp.com
Sales Inquires: sales@niccomp.com

www.niccomp.com

**ALUMINUM • TANTALUM • CERAMIC • FILM • RESISTORS • INDUCTORS • FERRITE CHIP BEADS
• THERMISTORS • VARISTORS • DIODES**

Continued from page 10

causes defects like cracking and curling. An adjustable air flow exhaust fan draws HEPA filtered air over the tape in a counter flow direction. As the atmosphere flows from the dry exit end of the tape casting machine oven to the front end where skin formation is the biggest problem, it slows and becomes saturated enough to prevent excessive drying. Microprocessor controls and sensors can be set to maintain a desired atmosphere saturation gradient across the drying oven.

After drying, the moving tape can be subjected to quality control checks, cut to length, trimmed and slit into narrower widths before being collected or rolled onto a product take-up spool (Fig. 2). The newest tape

Figure 2

casting machines incorporate the best technology available, such as automatic web tracking to align the tape precisely so take-up is smooth and seamless. Other "bells and whistles" can be added to meet particular product manufacturing needs.

Figure 3

Today, tape casting machines utilize the latest computer control technology (Fig. 3). Human machine interface (HMI) control systems deliver a simplified operator interface and graphic displays provide essential information and ease of control, leaving almost nothing to chance. These "what you see is what you get" (WYSWYG) systems (Fig. 4) are

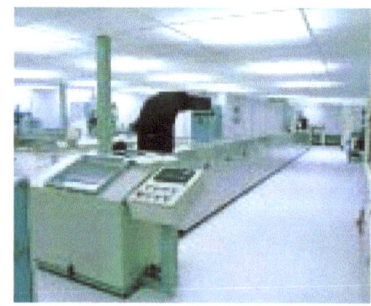

Figure 4

easy to use, shorten operator training time and dramatically increase productivity.

Although this technology was limited in the past, there should be no doubt that more recent developments will continue the march of progress in tape casting machine technology. Perhaps this could even lead to the "perfect" machine to make the "perfect" part every time.

HED International, Inc., is a leader in the design and manufacture of processing systems for technical ceramic and non-ceramic materials. For more information, visit the company website at www.hed.com. ❑

Featured Technical Paper

Automated Solutions for Passive Component Dicing

*Thomas Lieberenz and Joel Sigmund,
Disco Corporation*

One of the challenges in the manufacturing of passive components is how to singulate the individual components from the bulk format in which they are produced. The bulk format is either a substrate or a wafer and the material is usually ceramic. However, a wide variety of other materials such as ferrite, silicon, and compound materials may be used (Figure 1). There are several methods available for the

Figure 1: Diced fired ceramic.

singulation process, including dicing, laser cutting, scribe & break, routing, and punching. Each method has its advantages and drawbacks; the selection of the optimal process depends on such variables as material characteristics, substrate thickness, cut quality requirements, throughput, and precision.

Primarily due to its superior cut quality and accuracy, dicing is a commonly used process that provides flexible, high yield singulation for electronic components. Passive components such as ceramic capacitors, resistors, SAW (Surface Acoustic Wave) devices and oscillators are all manufactured with dicing equipment for this reason. Other singulation techniques such as scribe & break or punching are used largely because of their higher throughput rates. However, the increased speed of those processes present some drawbacks so they are typically only used for less demanding applications that do not require the precision and quality of dicing.

Reducing cost has been an effort in many high tech industries but nowhere else is the pressure to cut cost as significant as it is in the passive component industry. Dicing equipment suppliers, such as Disco Corp., have responded to this challenge by working closely with electronic component manufacturers to improve process efficiency and lower operating costs. This article briefly reviews the basics of dicing technology and discusses recent advances in "state of the art" dicing technology that have increased process throughput and efficiency to lower a component manufacturer's overall cost of ownership.

Dicing Process Overview

Dicing saws employ high speed, low vibration spindles to rotate a dicing blade up to 60,000 RPM (Figure 2). A dicing saw will also have a feed axis (X), index axis (Y), cut height axis (Z) and rotational axis (theta) to control cut position and depth. A substrate or wafer is mounted on either tape-and-frame or to another substrate (usually glass) to secure it during dicing. The dicing process requires the substrate be loaded and aligned to the saw axes either manually or automatically prior to cutting.

Figure 2: Cutting process of a ceramic substrate.

Continued on page 16

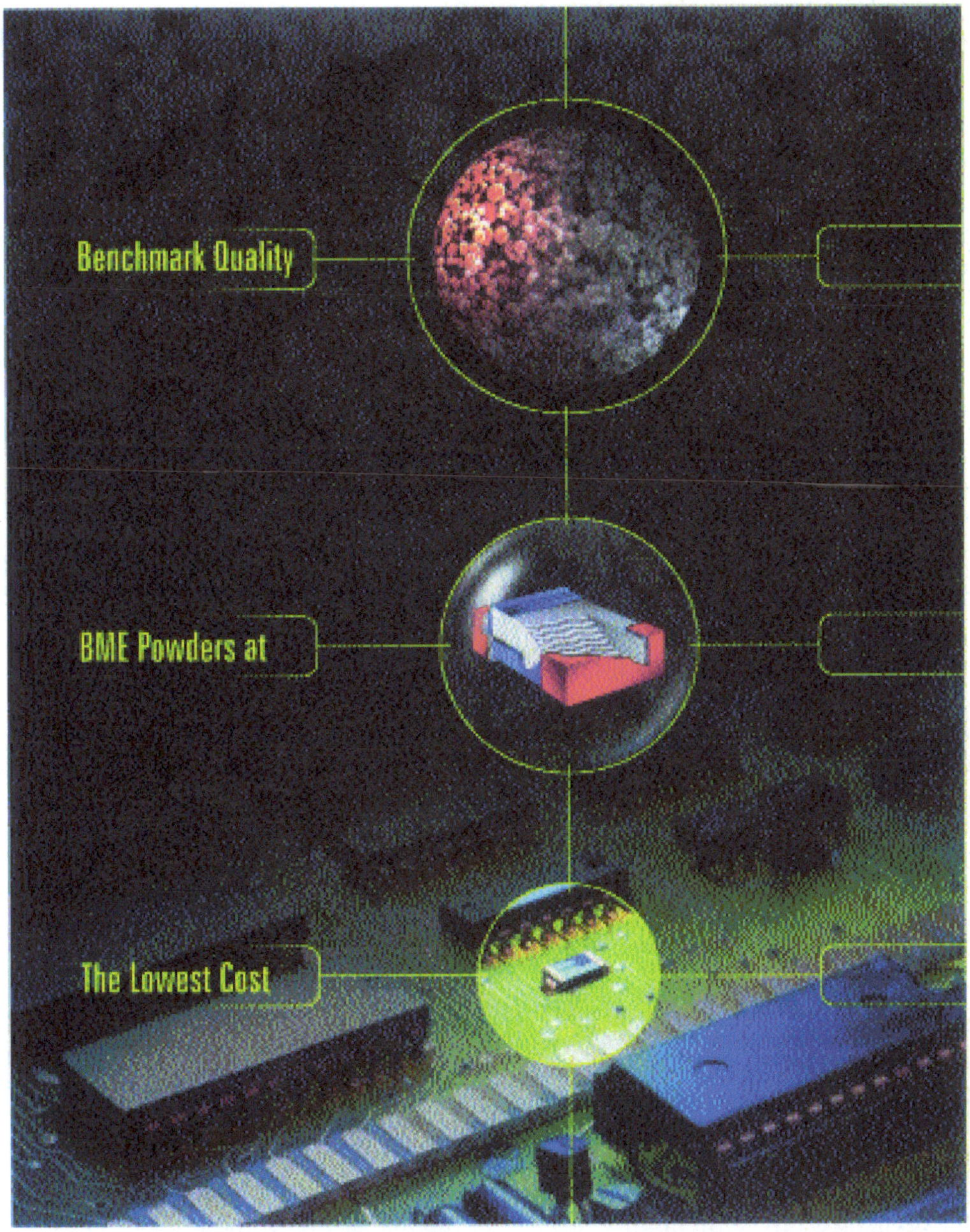

Benchmark Quality

BME Powders at

The Lowest Cost

MAKING YOUR TRANSFORMATION TO BASE METAL SMOOTHER

CEPC is in the business of producing fine nickel & copper powder for the manufacture of base metal multi-layered chip capacitors (BME-MLCC). With its high performance powders and technical expertise, CEPC is ready to team up with your professionals to make your transformation from precious metals to base metals smoother. Thinner electrodes, more layers, higher capacitance are now within reach with our <0.3 µm powders. In association with our partners H.C. Starck, we have established World class teams to serve the North American, Asian and European MLCC industry through our joint distribution and customer assistance centers located in the U.S.A., Japan & Germany. Using high purity feed materials and a proprietary vapor phase technology that allows the controlled nucleation and growth of particles, CEPC's production process offers the unique ability to tailor the powder properties to meet even the most stringent demands of BME-MLCC manufacturers. These properties include degree of crystallinity, spherical particle shape and a mean diameter ranging from 0.1 - 1.5 µm. The spherical and highly crystalline nature of CEPC's powders make them highly resistant to oxidation and shrinkage. For more information, please contact us.

3494 Ashby Rd. - St-Laurent - Québec - Canada - H4R 2C1 - Tel.: (514) 336-2888 - Fax: (514) 336-5059

Continued from page 14

Water is used during the cutting process to cool the blade and to wash debris out of the cutting area. After completion of the dicing process, singulated components are typically processed further and then placed into a transport medium, such as tape-and-reel.

Dicing blades consist of an abrasive (grit) such as diamond, that is held together by a bond material. Factors such as grit size, grit concentration and bond hardness greatly influence cutting results. Cutting results can be broken down into three categories: (1) quality; (2) precision; (3) efficiency. Cut quality refers to how the cut edge of the substrate looks. The dicing process will create chipping, cracking or burring mostly depending on the material type and thickness. Cut precision refers to the straightness, perpendicularity or shape of the kerf. Non-optimized blade formulation and process parameters are a common cause for inferior cut precision. Finally, cutting efficiency describes the throughput and blade wear. The faster a blade can cut, the better the throughput will be, but this will also impact the blade wear, quality, and precision. Process development is important to achieve balanced yet optimal results for each of these factors.

Selecting the right blade and process parameters is a complex task involving a number of variables. In order to identify the right combination, extensive experience and process knowledge is necessary. For example, improving cutting efficiency by increasing feed rate will generally result in a reduction of cut quality (increased chipping). One solution for maintaining cut quality while increasing feed rate is to raise the spindle RPM proportionately.

Blade cost has a direct impact on the cost of the final product, so component manufacturers tend to focus on increasing blade life (reducing wear) without adversely effecting cut quality, precision, and throughput. This is achieved by the right blade selection and by new blade development efforts. Blade wear is the result of the bond wearing away, releasing old, worn out grit and exposing new, sharp ones. The blade is in effect sharpening itself as it cuts. This "self sharpening" characteristic is important to dicing and makes it unique compared to knife blade-based singulation processes.

The electronic component industry uses a very broad range of materials that require a wide range of dicing solutions. Every dicing application has its unique challenge. For example, LTCC is often several millimeters thick when diced and may have excessive chipping and cracking when material glass content is high. The solution is a softer dicing blade bond that releases worn out diamonds sooner to improve its cutting ability. We can better understand the necessity of dicing process development by looking at the example of cutting quality optimization for SAW devices. SAW devices are made of relatively hard and brittle ceramics (Quartz, Lithium Tantalate or Lithium Niobate) that may also exhibit a high degree of chipping during dicing. Dicing coolant additives are employed to reduce chipping while maintaining throughput.

Application-specific process development is critical to achieve a dicing process optimized for low cost, high throughput and quality. Given the technical complexity and variety of materials within the components industry most component manufacturers rely heavily on the expertise and resources of dicing equipment suppliers to fill this need. Disco Corp. makes available its vast applications experience and offers this service free of charge to its customers at a number of applications labs around the world.

Multi-Blade Dicing

The most intuitive way to increase dicing process throughput is to cut with more than one blade (Figure 3). Gang blade assemblies are now employed in production with blade counts from 2 to 40 blades per spindle. This technique is employed in the singulation of numerous products such as ceramic capacitors, resistors and fuses. The number of blades to be used depends upon the hardness and thickness of material to cut, positional accuracy requirements and the power output of the dicing saw spindle.

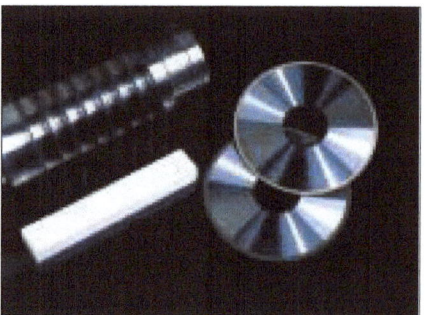

Figure 3: Gang blade assembly.

What is not intuitive is that the throughput of gang blade dicing is not linearly proportional to the number of blades used. When ganging 40 blades, for example, you may not realize a 40X increase in throughput as the feed rate may have to be reduced to drive that number of blades and set up time will certainly increase. Gang blades require longer set up time to balance the arbor assembly and adjust its cut height. Uneven blade wear or blade breakage leads to early replacement of the gang blade assembly and increased set up time, making blade selection critical to success.

Mechanical bearing spindles with higher torque are required if gang blades are used to cut hard and thick materials. Conventional dicing saws employ air bearing spindles that use air pressure as a medium to hold the rotating spindle in its shaft. Air bearings are virtually free of vibration but are generally not recommended for gang blade applications due to the higher loads of the arbor assembly. In contrast, mechanical spindles use ball bearings to hold the spindle accurately in its shaft even when high forces are applied.

For applications requiring high cut accuracy and process flexibility, dual spindle dicing saws are the machine of choice (Figure 4). Single blade dicing has been particularly effective at minimizing dimensional variability of the final product. Each spindle is equipped with a single

Figure 4: Cutting process of a silicon substrate with a dual spindle dicing machine.

blade that moves independently to ensure desired placement of the cut. Blade height is controlled automatically and independently. There is an up to 2X increase in throughput with the dual spindle saw and the problems associated with multiple blades on an arbor are eliminated. Dual spindle dicing saws combine the benefit of using more than one blade for higher throughput and the high accuracy and low set up time of a single spindle configuration. For example, thin film on silicon passives can be diced with a dual dicing machine to achieve higher throughput while maintaining the high accuracy needed to allow small margins between the passive components.

Automatic Alignment

Alignment refers to positioning the substrate so the blade cuts at the intended position. Manual alignment requires the operator to align the substrate using a microscope. Automatic alignment utilizes a vision system to find and recognize alignment targets on the substrate so the machine can adjust the blade and substrate position automatically to achieve the desired cut position. Automatic alignment removes the operator from the alignment process and provides the capability to handle uncontrollable variations in the manufacturer's substrate. The result is a process with higher throughput and yield.

Removing the human element from the process guarantees a faster, more repeatable process. Manual alignment typically takes one to two minutes, depending largely on the skill of the operator. Automatic alignment times vary but usually do not exceed twenty seconds. In processes with very short cycle times, auto alignment can be a huge improvement to overall process throughput.

Automatic alignment can also continuously adjust cut position to compensate for process variations in the substrate. Such adjustments would not be feasible for manual operation. For example, ceramic

Let our Amitron unit's world-class LTCC service get you a prototype ASAP: Send for our new, free design guide!

Need a multilayer network for a high-density, high-frequency application? Looking at a complex design full of interconnects? Our Amitron subsidiary will do the job right — and right quick!
- $1 million in new, world-class LTCC equipment manned by an experienced staff
- 200 mm format capability
- Established DuPont, Ferro & Heraeus materials and systems
- Stringent process control guarantees performance
- For added integration, we also offer:
  – High-performance, advanced etched photolithography conductor processing
  – Passive element tuning by YAG lasers
  – Gold and solder plating for robust manufacturing
  – 100% continuity testing and brazing capabilities
  – Plus world-renowned microwave-circuit design and testing, by Anaren

To learn more about Amitron LTCC — call the number below or e-mail ltcc@anaren.com

800-411-6596 > www.anaren.com
In Europe, call 44-2392-232392
ISO 9001 certified
Visa/MasterCard accepted
(except in Europe)

substrates experience varying degrees of shrinkage during the firing process. During firing, the component pattern will move differently depending on a number of factors, but the result is an uneven spacing of the components that a fixed index dicing process cannot handle. The solution is to have the dicing saw detect the actual components instead of just one alignment target. As you can imagine, having an operator align the substrate for every cut line is impractical. Therefore, automatic alignment would be recommended.

As most component manufacturers use manual dicing equipment requiring manual alignment, adequate targets for automatic alignment may not exist. The key for the auto alignment process is the ability of the alignment system to recognize the target. Recognizing the target is easy for very accurate and fine patterns such as photolithography patterns on a silicon wafer. However, many components consist of thick films on ceramic that have poorly defined edges and are often not placed accurately, resulting in inaccurate cuts. Special software algorithms are used in these cases to compensate for the inaccurate alignment target.

Special alignment processes must be developed for components that have no patterns visible at the top side. For example, MLCC (Multi Layered Ceramic Capacitor) substrates have a blank ceramic layer on the top. The process that Disco Corporation recommends is to expose electrode patterns by grooving a v-shape at the sides of the top layer with a bevel-shaped blade. The exposed, visible pattern is then used to align the cutting streets.

Automatic Handling

Manual dicing equipment is most common within the electronic components industry, however, the trend today is to automate this process to reduce labor costs, increase yield, and increase machine utilization. As substrate cycle time (cutting time per substrate) decreases with the introduction of multiple blades and auto alignment, it becomes more difficult for an operator to keep up with the machine. A machine may sit idle waiting for an operator to unload/load material, rendering the process less efficient. At one company, the introduction of automatic handling increased machine utilization from 40% to 70% and decreased the number of operators required in the dicing area. The challenge the dicing equipment manufacturers now face is how to handle such a broad array of substrate configurations.

Fully automatic dicers are typically loaded with a cassette that holds all the substrates, which are mounted to tape and frame. The dicer takes the substrate out of the cassette and transfers it to the chuck table using a transfer arm that pulls vacuum on the metal frame of the tape. After dicing, the substrate is transported to a washing station which removes debris from the substrate by spinning the substrate and spraying water on it. After the substrate is dried, the handling system moves it back into the cassette. Fully automatic manufacturing sites may employ robot systems that move the cassette from machine to machine. In many cases, the cassette is moved from the taping machine to the dicing machine and then to the pick-and-place machine.

In some cases it is possible to use a special tape that is hard enough to hold the substrate without a frame. Instead of holding on to the frame, a special handling unit holds the substrate directly. This handling unit uses a plate that applies vacuum in several areas to hold the substrate. An example of a component where this handling system is used is for MLCCs (Figure 5). The diced MLCC substrate can be transported to the firing stage without the need to release it from tape and frame.

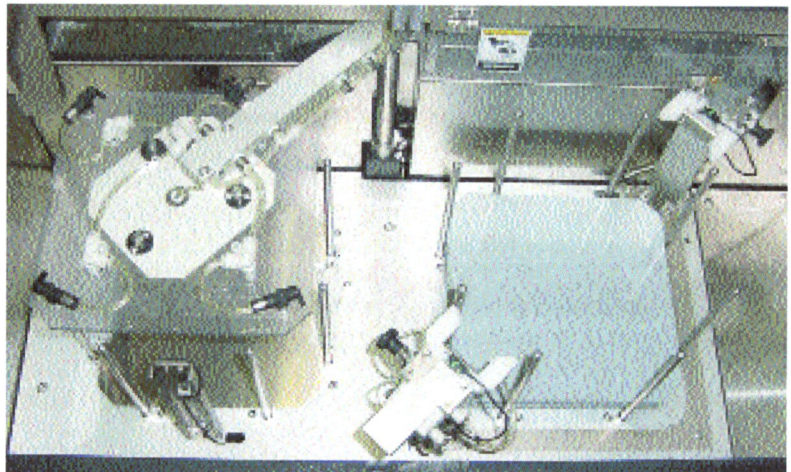

Figure 5: Special handling system of fully automatic dicing saw for frame-less ceramic substrates.

Conclusion

The industry trend of producing high quality components at a lower cost forces electronic component manufacturers to continuously improve the efficiency of their operations. Due to the many variables involved, the dicing of electronic components is technically complex. Defining an optimized process requires matching the correct blade, machine configuration and level of automation specific to each component manufacturer's needs. As products, materials, and requirements continue to change rapidly, an important resource of the component manufacturer is the dicing equipment supplier's ability to provide ongoing service and applications support, while driving technology advancements in equipment and blades to further improve processes. ❏

Cheers!

The Oberon Group announces its formal association with **The Paumanok Group** to combine their respective financial and strategic expertise for the passive electronic component industry.

The Oberon Group is a seasoned team of investment bankers who have come together for the express purpose of providing the highest level of M&A, private placement and financial advisory services to small and midsize businesses.

The Paumanok Group is the world's largest supplier of market research and intelligence services to the passive electronic component industry, and has represented many of the leading industry players for over a decade.

RSVP

THE **OBERON** GROUP, LLC	The Paumanok Group
Adam Breslawsky	Dennis Zogbi
Managing Director	President
(212) 798-1473	(919) 468-0384

Continued from page 4

in demand for alternative dielectrics, with emphasis upon base metal electrode MLCC. Still, by the fourth quarter of 2002, all companies, save AVX vs. Cabot, had settled their lawsuits and agreed to work together (who can argue with the fact that Cabot emerged victorious in most of the disputes to date–who else can guarantee they will earn more than $550 million in tantalum powder sales over the next three years?).

Also in tantalum capacitors at the end of 2002 we began to see signs of new interest in the dielectric in Japan. This interest was primarily and ironically in the smaller P, J and new 1005 case size parts, typically for applications in digital video circuits for cell phones and cameras. This is quite interesting because the capacitance values in these small case size tantalum chips are right in the area where the high capacitance MLCC products have their greatest capacity to produce– between 1 and 10 µF.

In DC film capacitors, the situation was dire because not only was demand adversely affected by the global economy, but the specialty film capacitors for AC & Pulse and interference suppression (X and Y) were under constant pressure from value-added ceramic capacitors. Moreover, the movement away from CRT monitors toward flat panel displays seriously affected demand for AC & Pulse film capacitors used in degaussing circuits.

In aluminum capacitors, there was serious competition between the Japanese and the Taiwanese for control of the mass produced products used in consumer A/V equipment and computer motherboards. Many of the traditional top tier manufacturers in aluminum electrolytics continued their development work on value-added products, with emphasis upon low ESR polymer aluminum configurations and chip types in general. Electrolytic manufacturers in Japan also continued extending into the value-added double layer carbon capacitor arena.

In chip resistors, the market remained especially depressed in terms of both volume and value. This was interesting to me because we saw a 20% increase in MLCC unit sales between April and July of 2002 to the Asian countries outside of Japan (with emphasis upon Taiwan). However, we saw no correlating increase in chip resistor sales during the same time period, and MLCC and Chip resistor unit sales usually track each other. This has been the subject of some debate in the Asian community. Some major suppliers believe Taiwanese motherboard manufacturers overestimated demand in the computer industry and bought more MLCC than they needed (thus there may be stocks in Taiwan). Others thought that excess inventories of chip resistors, left over from 2000, were higher than their MLCC counterparts, so in fact all parts were consumed and no excess remains in the region (optimists vs. pessimists).

The forecasts produced by Paumanok for 2003 suggest more of the same, with emphasis upon continued price erosion. However, we expect unit shipments to grow, and we expect the value of shipments to grow as well, but only slightly. Typically the third year in the cycle represents the most difficult year, when small companies competing in the mass market go out of business. Only the larger companies, with massive economies of scale, or the smaller businesses that produce value-added and application-specific passive components are able to survive. Paumanok predicts a gradual recovery in the industry, one that exhibits a combination of up and down quarters (better than 2002), but full recovery will take time. We believe there will be no pronounced turn until the second half of 2004, an upturn that will be pronounced and last until October of 2005.

In the meantime, only the smart companies will survive. In mass commercial passive component markets, it will be those companies that have large account status at the larger EMS companies, and those with access to global distribution. It will also be those companies with access to China (i.e. the right China reps). It will also be those companies with direct customers in wireless, LCD displays, and digital video. For the smaller specialty passive component producers, obviously, those with positions in defense electronics, medical electronics and specialty industrial products (i.e. downhole pump, mining electronics and undersea cable repeaters) will also fare well until the global markets recover in 2004. ❑

If passive components are your business, there is one event you don't want to miss:

The 23rd Capacitor and Resistor Technology Symposium

CARTS 2003

March 29 – April 4, 2003
Chaparral Suites Resort
Scottsdale, Arizona

Capacitors

Resistors

Integrated Passives

Magnetics

Filters

For more information visit
www.ec-central.org/conference/CARTS2003

Materials Research Furnaces, Inc.

Introduction:

Materials Research Furnaces, Inc. was founded in 1990, by a group of highly trained and experienced engineers and technical personnel from within the vacuum & high temperature furnace field. **MRF Inc.** was established to answer the challenge of the Research & Development community to produce the finest, high temperature, high vacuum and controlled atmosphere furnaces in the industry.

MRF BUILDING

TANTALUM CAPACITOR SINTERING FURNACE

MRF Inc. has supplied new furnace systems and replacement parts to Universities, National Laboratories and to private industries around the globe. A large part of **MRF's Inc.** services is providing parts and services for other furnace manufacturers' systems. **MRF Inc.** has over 14,200 square feet of space for manufacturing, assembly, engineering, sales and after market support.

Materials Research Furnaces, Inc. takes great care in providing you, the customer, a quality product that is both reliable, simple to operate and user friendly. The operation of any **MRF Inc.** system can be mastered in just a few hours. **M R F I n** furnace systems will complement any laboratory or manufacturing facility.

Products & Furnace Outline:

MRF Inc. produces a wide range of furnace for almost every application. Our furnaces range in temperature from 600 degrees Celsius to 3000 degrees Celsius. Our vacuum furnaces are designed to the 10-9 torr range. These furnaces can be used in a variety of inert gasses and other volatile gasses such as hydrogen and methane. Some of **MRF Inc's** furnaces are:

Continuous Belt Furnace:	**Top & Bottom Loading Furnaces:**
Hot Pressing 1 through 100 Ton:	**Front Loading Batch Sintering:**
Physical Testing:	**Graphite Tube Furnaces:**
Arc Melting Furnace:	**Crystal Growing Furnaces:**
Muffle Tube Furnace:	**N2 BME Furnaces:**

For more information:
Contact **Materials Research Furnaces, Inc.,**
Suncook Business Park; Rt. 28 & Lavoie Drive
Suncook, NH 03275: Attn: Daniel J. Leary, SM
Phone: (603) 485-2394 Fax: (603) 485-2395;
E-mail: mrf@interserv.com
WEB SITE: www.mrf-furnaces.com

Reps in Europe & overseas representation:
In the European Union contact:
Instron SFL
Severn Furnaces, Ltd.
Mr. Stephen Horrex
Brunel Way
Thornbury, Bristol BS 35 3UR, UK
Phone: 1454-414600; Fax: 1454-413277;
E-mail stephen_horrex@sfl.instron.com
All other overseas contact MRF directly.

TANTALUM CAPACITOR SINTERING FURNACE

"TODAY'S FURNACES FOR TOMORROW'S TECHNOLOGIES"

Featured Technical Paper

Improvements in Electrode Technologies

*Ann Pogue and Richard Stephenson,
Ann Pogue & Associates, Inc.*

Abstract

Increased performance requirements, smaller package sizes and less expensive material composition are the ongoing technology strategies for Multi-Layer (ML) component makers. To achieve these goals, improved formulation of ceramics, metals and processing are being used. Electrodes with high silver content AgPd powders, pure silver or base metal powders (nickel and copper) will be discussed.

This paper compares three technologies to achieve highly dispersed electrode powders in pastes, including fabrication of powders (precipitation, spray pyrolysis, etc.), mechanical deagglomeration and chemical modifications. The effects of electrode powder particle size, size distribution and interfacial surface properties are described, as well as methods to improve reactivity when using nano-powders.

1. Powder Manufacture

A wide variety of methods are used to make metal powders for electrodes in ML components. Although

various production methods are capable of making metal powders, the final material properties do not equally compare. Naturally, the physical property of the metal powder has an impact on the process of fabricating electrode layers in ML components. The impact is manifested in different ways; the performance when formulated into a paste, reactivity of the paste during the burnout cycle and the thermal behavior of the metal during the sintering stages of firing. Features of metal powders that are affected by route of manufacture include particle morphology, particle size distribution, surface reactivity, oxidation behavior and functional chemical affinity.

The most significant metal powder manufacturing technology is chemical precipitation. Recently, other powder manufacturing technologies have become prevalent, but remain secondary to chemical precipitation. The secondary technologies include, but are not limited to, spray pyrolysis, vapor deposition and cryogenic milling.

The selection of manufacturing method is dependent on the final composition of the powder, ease of production and process cost. Process flexibility and robustness is also a consideration. For these reasons, chemical precipitation is widely used. Many ML component electrode compositions remain precious metals, typically silver-palladium, with the trend moving toward pure silver. In addition, base metal electrodes of nickel are routinely being made on the industrial scale (billions of units per month) and full commercialization using copper powder is well underway.

Chemical precipitation allows the metal powder manufacturer to control multiple aspects of the product. Reactor medium (usually water), pH, temperature, reducing agent, agitation dynamics and reaction time are some of the process control points to influence the final powder properties. Production quantities can be quite large (50 – 500 kg and more). Typically, an organic dispersant is selected to add control to the particle size, shape and distribution. In addition, careful selection of a dispersant system facilitates not only the production of the metal powder, but also the making of pastes for the customer. Difficulty arises in choosing a dispersant system since most of the industrially important chemical precipitation methods are done in water and the final metal powder must be dispersed into an organic phase (paste). Failure to control the particle properties during precipitation can result in moderate to highly agglomerated powder. This in turn adds additional cost to the powder manufacturer in additional processing and also inhibits the manufacture of very thin, high layer count ML parts for the final component maker. Figure 1 shows two SEM images of chemically precipitated 95%Ag5%Pd powders by different manufacturers. Both powders are used in the same application, however,

distinct differences can be seen between them. Figure 1a shows a deagglomerated powder whereas Figure 1b shows clear agglomeration of particles, especially the larger particles.

Neither of these powders is particularly superior. The "A" powder, although deagglomerated, contains a large population of finer particle sizes and has an overall wider particle size distribution. Fine particles have the effect of increasing paste viscosity as well as acting to agglomerate to each other when milled into an organic phase. The "B" powder contains a much higher population of larger particles, but also agglomerates of these particles. Clearly seen in the "B" image of Figure 1 is a classic "dumbbell" as well as "necked" and "chained" agglomerates. These types of agglomerates are very hard to break up so roll milling typically produces "flakes" during milling. Figure 2 shows the SEM image of a 95%Ag5%Pd chemically precipitated powder exhibiting good control of particle morphology, size and state of dispersion.

Measurements can be made of metal powder to compare how easily dispersed they will be in a given medium. Most modern particle sizing instruments are capable of using different media and also provide inline sonic or ultrasonic probes to aid in dispersing powders in

Figure 1: SEM images of 95%Ag5%Pd by different manufactures by precipitation.

Figure 2: SEM image of 95%Ag5%Pd by precipitation.

MarketEye is a free information service for buyers, engineers and managers who are concerned with passive, interconnect and electromechanical components. In addition to industry experts' analyses and insights on market trend information, MarketEye now provides new features using current, comprehensive data.

- **Lead-Times/Lead-Time Trends** — Information on more than 100 product families from TTI's industry-leading data base.
- **Obsoleted Parts** — Listing of obsoleted components from all TTI manufacturers.
- **Manufacturer Holds** — Listing of Commercial and Military/E-Rel products placed on hold by TTI manufacturers.

Go to ttiinc.com/marketeye and see how MarketEye can work for you.

www.ttiinc.com

dispersion media during PSD measurements. Figure 3 compares the PSD results obtained on the powders in Figure 1 when different sonic energy is applied during PSD measurement. A clear difference is seen in the strength of agglomeration when comparing the powder from Figure 3a and Figure 3b. Although the "A" powder appears to have a better PSD trace at 25% power, it also shows it is more difficult to disperse with increasing power as seen by the smaller changes in the PSD trace above PSD D90. The "B" powder, although larger in particle size and agglomerated, was far easier to disperse when tested in the same dispersion media and under identical conditions.

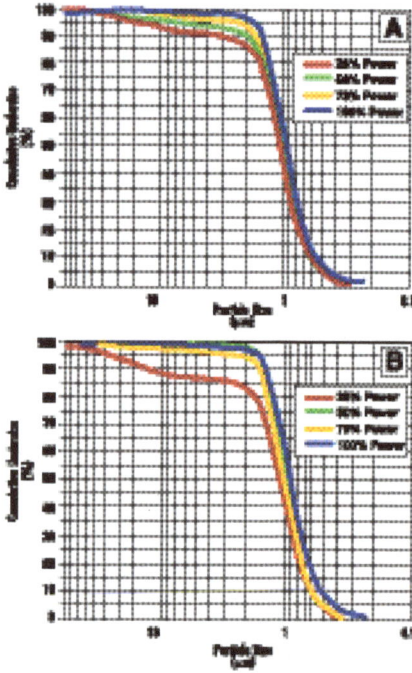

Figure 3: Particle size distributions obtained from powders in Figure 1 by applying different strengths of sonic energy.

2. Mechanical Deagglomeration

Limitations to the dielectric layer thickness of multi-layer parts demand better performing electrode materials to achieve higher layer counts, smaller part size and improved component capability. In order to achieve this, very thin electrode laydown must be achieved. Making the metal particles smaller would seem like a first approach to accomplish this, however, too small of particle in the electrode layer becomes very reactive during burnout and firing, causing minor and massive defects to the final part. Figure 4 shows the relationship between particle size and specific surface area for various compositions of silver palladium and nickel powders. A metal particle of 0.5 micron diameter is expected to

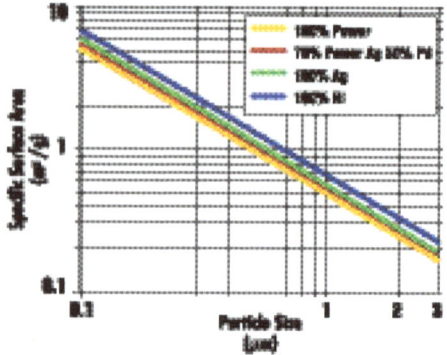

Figure 4: Theoretical relationship between metal particle size and specific surface area.

have a surface area near 1 m²/g. However, most metal powders exhibit an excess surface area over the equivalent spherical diameter, sometimes as much as 50%. This is due in part to the method of manufacture. Chemical precipitation, for example, generally produces metal particles with very small crystallite sizes; on the order of 15 – 50 nm. This means that an individual metal particle of pure silver, for example, will have numerous grain boundaries. The excess surface area also arises from the particle not being a perfect sphere. Pores on the surface contribute significantly to the excess surface area of metal powders.

Another reason for deviation of specific surface area from expected surface area is due to chemical properties of the metals. Nickel is very reactive to atmospheric oxygen and moisture as the particle size decreases. Surface oxides on nickel and other metals, contribute to higher surface areas.

There are several methods to mechanically separate metal particles. Most of these are post-processing methods, that is, after the particles have formed and may or may not be agglomerated. It is at this stage that energy is applied in the form of mechanical action to separate agglomerates. This may be done by the paste manufac-

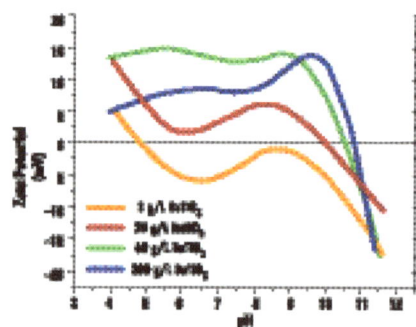

Figure 5: Plot of zeta-potential against pH for $BaTiO_3$ in aqueous media at different $BaTiO_3$ loadings.

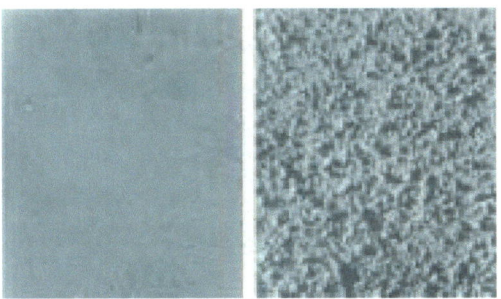

Figure 6: Low angle light images of well dispersed and poorly dispersed 95%Ag5%Pd powders (from different vendors).

turer, the powder supplier, or both, to modify the surface of the powders so they are tailored for each usage.

One method is sand milling. This process allows high energy to separate the agglomerates and allows tailored coating to be placed on the particles to prevent agglomeration either with other metal particles, inorganic fillers or around resin or plasticizer particles. This type of process typically is done to balance the electrical charges in the metal/solvent/resin system. Figure 5 shows the surface charge behavior of barium titanate in aqueous media at different loadings. Adding other components to a formulation creates greater surface charge complexity. Figure 6 shows the effects of using 95%Ag 5%Pd from different vendors in otherwise identical paste formulations.

Equipment can be used to find the best coating for the inorganic particles when combined with the actual resins and plasticizers used for a customer's paste system. Optimally, if this balance is achieved, paste can be made with high speed dispersion and eliminate milling steps. This process needs to be done with both the metal and the filler material to allow good mixing of the paste that is maintained with time (Figures 7 and 8).

A failure to balance the system is frequently seen as a "white" separation along the sides of the paste container or on the top or bottom of the storage jar.

The classic methods of electrode manufacture have been three roll milling or ball milling. With three roll milling, the agglomerates are separated by high shear. If the metal powder contains large agglomerates, these may be 'coined' by the milling process rather than dispersed as in Figure 9. For this reason, it is very important to fully mix the paste in high-energy mixers before milling. Roll milling works well for electrodes that are printed moderately thick, however roll mills usually leave some agglomerates that will cause difficulties with thin dielectric layers. When using roll mills to produce thin electrodes, it is very necessary to pre-treat the powders, such as using sand milling mentioned above, to the inorganic particles pre-coated by an organic that is compatible with the resin system used. Without pre-coating, three roll milling usually produces some agglomerates, either through "coining" or through agglomeration of fines around either the resin or the ceramic additions.

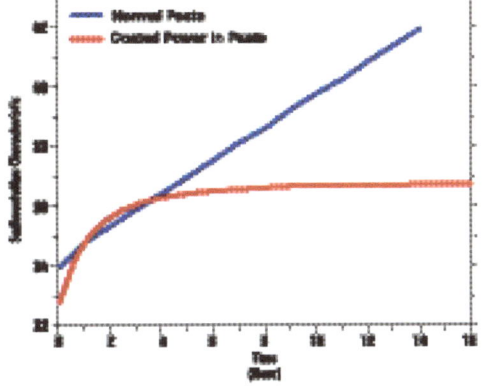

Figure 7: Paste dispersion stability traces for 95%Ag5%Pd.

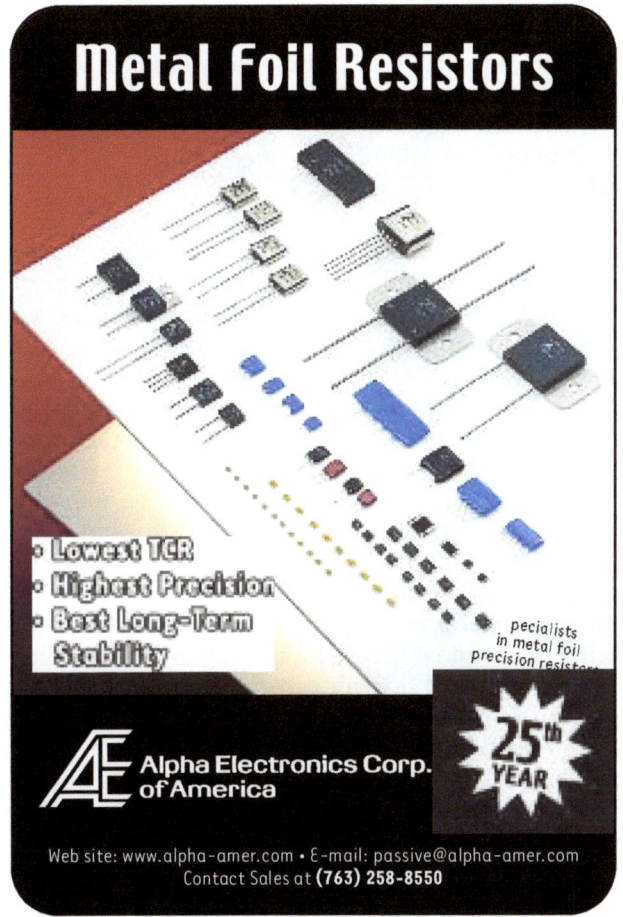

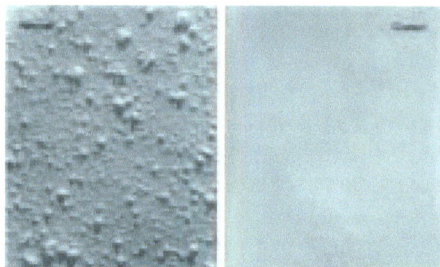

Figure 8: Low angle light images of the pastes samples of Figure 6 (paste dispersion stability).

Figure 9: Typical results of "coining" metal particles before (left image) and after (right image).

The ball milling process mixes the particles by shear and media impact. This technique works best when the particles have been coated. In this way, the electrical balance of the constituents is maintained to provide optimal dispersion when being mechanically acted upon in the resin system.

Newer processes for making electrode pastes include ultrasonic and liquid jet milling. Both of these processes will break apart "soft" and moderately hard agglomerates. However, if the electrical charge balance is not accounted for, re-agglomeration can occur after the dispersion treatment. Other problems have also been noted when too much energy is applied to the system being dispersed; fine inorganic particles react with the organic phase. This effect can be difficult to discern until late in the fabrication process of a microelectronic part. Figure 10 shows the effects of a paste sample with and without fine material present, resulting in a fully fired part and a part with moderate and severe defects. The reactivity of fine material present in the paste results in difficulties during burn out causing delaminations due to evolution of vapors after the organic constituents of the paste and ceramic tape have decomposed.

Sand milling the final paste is becoming much more common in the microelectronic component industry. The older methods of roll milling and ball milling arose from the paint industry, whereas sand milling was developed more from ink industries. As the microelectronic component manufacturers continue to strive

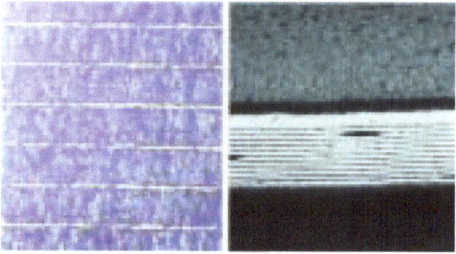

Figure 10: Image of good fired character (left image) and moderate and massive defects (right image).

for thinner and high speed printing, the dispersing technology used to accomplish this will become more like what is used in metallic high speed inks and ink jet systems.

3. Chemical Modification
<u>Dispersion</u>

When the metal powder manufacturer can coat the metal powders for dispersion – and the paste maker can coat the additives and balance the electrical charges with the resin system – then a well dispersed paste capable of forming a thin continuous layer can be made. Dispersion characteristics can be greatly improved by modifying the surfaces of the metal particles comprising the electrode layers. Since the electrodes are formed from applying an organic phase to alternating layers of dielectric, the goal is to ensure an even distribution of metal particles in the organic phase and to maintain separation of each metal particle during the organic phase laydown. It is also critical that during burnout of the organic phase the metal particles are close enough to form a connected conductive sheet in the final product after the firing cycle. One additional desired feature is to fabricate a paste that has long term stability.

There are three principal approaches to achieve the chemically modified surfaces of metal particles. The first is through the method of manufacture. A stabilizing coating can be applied that not only aids in manufacturing and handling, but also preserves good dispersion characteristics when provided to paste manufacturers. However, in this case, the powder manufacturing method must not produce agglomerated material. Otherwise, these agglomerates would be carried to the end use in paste manufacture.

The second approach is to de-agglomerate the metal powder sample by some means, such as in section 2 of this article; by mechanical methods. Here again, careful use of technique is required so that the desirable properties of the powder are retained. Frequently in mechanical deagglomeration methods, over-aggressive applications cause deformation of the metal particles.

In sensitive powders, such as the high silver bearing silver palladium alloys or pure silver, too much mechanical energy can coin and flake the powder, rendering it virtually useless in the intended application.

Reactivity

As the industry evolves to the thinner layers containing finer particles of both ceramics and metals – the systems become much more reactive due to the higher energy in the small particles. These systems will sinter at lower temperatures and react with chemistries (functional groups) that were insignificant when the particles in the system were larger with lower surface area. For some systems it would appear that it is necessary to coat the metal particles with a substance that provides an ion such as barium or zirconium to lower the reactivity of the metal to the modifiers of the ceramic system. This becomes more important when working with hydrothermal and sol-gel type systems that have not been calcined and contain reactive modifiers. The reactivity must be selective so that ceramic reactions are preserved and do not interfere with the metal electrode or degrade the ceramic-electrode interface. Figure 11 shows the effects of adding an aluminum bearing compound and a silicon bearing compound to the surface of a copper powder. Although the expansion–shrinkage behavior can be influenced by the amount of material coated on the surface of the electrode powder, the choice of coating material will depend on how it modifies the thermal characteristics of the electrode, how it affects the electrode-dielectric interface, and the final electrical properties of the layered structure.

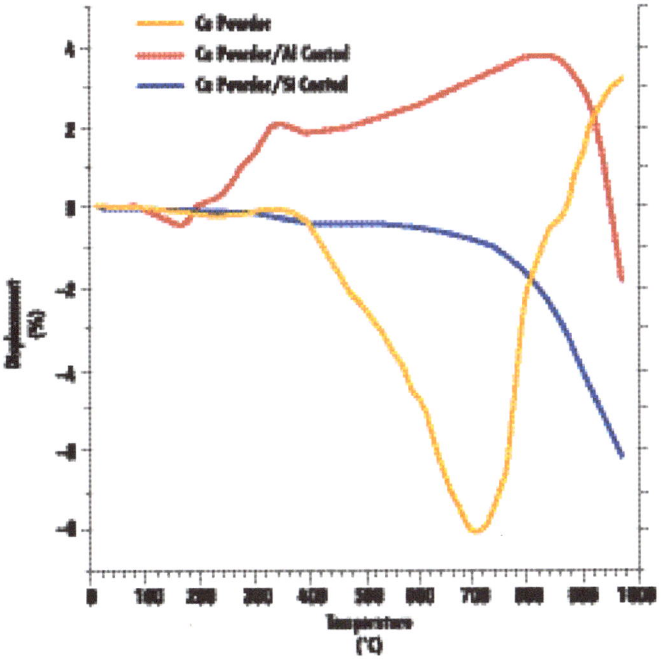

Other aspects of reactivity can be controlled by engineering the metal powder. For example, as the microelectronics industry develops high silver content electrodes, the problems of silver diffusion and evaporation become more difficult to control. One solution is to manufacture an electrode powder with a desired composition, but control the distribution of metal in each particle. For example, in a 95%Ag5%Pd electrode powder, the core of the powder can be manufactured to be 100% silver. Across the radius of a given particle, the concentration of palladium increases gradually so that at the surface of the powder the composition appears to be nearly pure palladium (Figure 12). The composition of the particle remains 95%Ag5%Pd, but the surface chemistry of the powder behaves like palladium.

Some of the reactivity control methods can be designed so as not to manifest dramatic changes throughout a manufacturing process. In the example of

Figure 11. Change in thermal characteristics by chemical surface modifications.

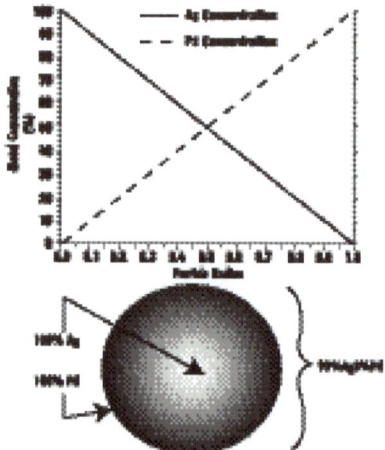

Figure 12: Schematic of a gradient electrode particle with a rich palladium surface. The metal gradient can be controlled for any composition and with any desired profile.

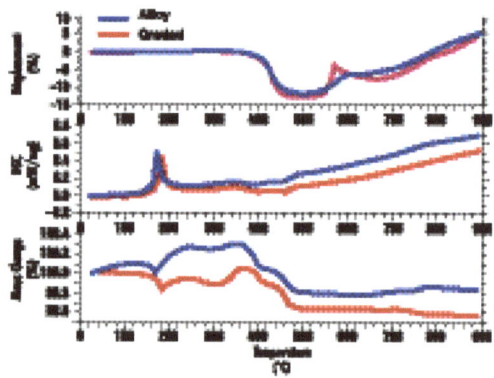

Figure 13: Thermal characteristics of 95%Ag5%Pd in air at 10°Cmin^{-1} of an alloyed powder (solid line) and a gradient powder (dot line).

Figure 12, 95%Ag5%Pd, the thermal characteristics closely match a metal electrode powder that is fully alloyed (Figure 13).

Shrinkage

As the metal particles become smaller, sintering and shrinkage rates are affected. There is a limit to oxide fillers that can be added without affecting the conductivity of the electrodes. Shrinkage can be modified by changing the crystallite size or by adding an inorganic coating that can delay sintering of particles without compromising the final desired electrical properties or electrode composition. Composite powders can also be used instead of admixtures, such as in the construction of the electrode layers by using a metal coated dielectric particle rather than adding fine ceramic material to the metal electrode powders (Figure 14). The composite system has several advantages over the admixture system.

Figure 14: Schematic representation of a metal particle (circle) and dielectric particle (irregular shape) system as an A) admixture, and B) composite material.

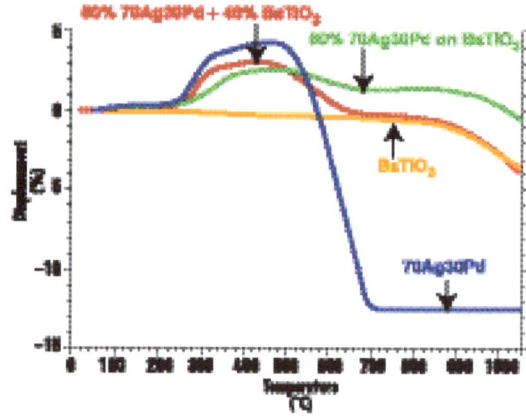

Figure 15: Expansion–shrinkage characteristics of BaTiO$_3$, 70%Ag30%Pd, a 60% metal + 40% dielectric admixture, and a composite powder of 60% metal and 40% dielectric.

The composite system ensures good electrical continuity since the surface of the composite powder is metal, whereas the admixture possesses discrete regions of pure dielectric. Both systems would appear to provide equivalent expansion-shrinkage control when equivalent amounts of each component are used to make the electrode layer. However, Figure 15 shows the comparison of the thermal characteristics of a pure metal system, an admixture of metal and ceramic, and a composite electrode powder.

Future Work

One area being developed is making the electrode "wet" the ceramic layer, thereby eliminating the gap found in electrode layers when ML components are frozen and split.

Conclusion

Much information can be gained to improve the stability of electronic pastes. Work can be done by the manufacturers in coating the powders and by paste makers in changing the balance of the surfaces to give

Technical Paper

a more stable and less reactive electrode. One can also control wetting and conductivity of the final products through the surface chemistry.

The surface chemistry needed to produce well dispersed, stable, fired electrodes becomes more important in non-fired systems. The surface of the conductive particles must link well with the function groups of the resin and modifiers to produce good stability of both paste and cured systems.

It is important to control the surfaces of the particles in the systems to gain high quality products as the particle sizes are decreasing and the layers in the components are becoming thinner. ❑

This article is based on a technical paper presented at CARTS-Europe 2002.

Letter From ECA

Continued from page 9

this model serves its constituencies. But, it fails to provide the links necessary to create equilibrium outside its sphere of influence. As a result, the supply and demand equations typically used to forecast electronic component movement have little impact on each other and shortages or oversupply occurs among different elements.

Using the flow wheel approach, knowledge of the supply and demand within each element impacts the speed and direction of the entire system. Just as car wheels can be made to rotate more efficiently or respond quickly to road changes, so can the electronics flow wheel. If a delay in one area of the wheel can be anticipated, then adjustments can be made and the situation can be moderated, keeping the industry on course.

ECA is already working with several organizations representing participants in the flow wheel. In addition to materials suppliers, ECA is working with NEDA and ERA to better understand the roles and responsibilities of the distribution sector. Through EIA, ECA is continuing to leverage direct links to counterparts in the consumer electronics and telecommunications industries. And, within a few months, ECA expects to develop a formal organization to address EMS and CEMs.

ECA's initiatives in 2003 will give every segment of the industry – materials suppliers, manufacturers, distributors, manufacturer's reps, OEMs and users – a much clearer picture of what is happening with the overall flow of supply and demand. Perhaps for the first time, the electronic components industry will be able to quickly understand what is happening within its complex system of interdependent groups, and have the collective understanding to quickly adjust. ❑

— Bob Willis is president of ECA, the electronic components sector of the Electronic Industries Alliance (EIA). He can be reached at robertw@eia.org.

Newsmakers

Vishay Introduces Miniaturized, Metallized, Polypropylene Film Capacitor

Vishay Intertechnology, Inc., announced the release of a miniaturized, metallized, polypropylene film capacitor designed to perform in high-voltage, high-current, and high-pulse operations at high frequencies.

Specified for broad voltage and capacitance ranges, the Vishay Roederstein MKP1841-M is suitable for a variety of applications including S-correction and flyback tuning, input and output filtering in SPS designs, and protection circuits in switchmode power supplies, snubbers, and electronic ballast circuits.

With competitive pricing and worldwide availability, the MKP1841-M offers designers a small, reliable, high-performing, and readily available capacitor for end products including TV sets and monitors, power supplies, and energy-saving lamps.

The new capacitor is available in small sizes, ranging from 4.0mm wide by 10.0mm long with a 9.0mm height up to 18.0mm by 41.5mm with a 32.5mm height, depending on capacitance and voltage ratings. Lead-spacing options range from 7.5mm to 37.5mm.

The MKP1841-M features 25 capacitance options in a range from 470pF to 4.7μF, capacitance tolerances of ±5%, ±10%, and ±20%, and voltage options of 250 VDC, 400 VDC, 630 VDC, 1000 VDC, 1600 VDC, and 2000 VDC. The device's maximum pulse rise time ranges from 133 V/μs up to 9610 V/μs, depending on lead spacing (PCM) and voltage rating.

Packaged in a flame-retardant plastic case with an epoxy resin seal, the MKP1841-M features exceptional self-healing capabilities, environmentally friendly lead (Pb)-free schooping layers and terminals, and vacuum-deposited aluminum electrodes. The MKP1841-M provides high reliability with an operational life of >300,000 hours, and a failure rate of <2 FIT at 40°C. The capacitor, which is rated for an operating temperature range of –55°C to +100°C, meets IEC test classification 55/100/56, according to IEC 60068, and is suitable for use with any other product from Vishay.

Samples and production quantities of the MKP 1841-M are available now, with 10- to 12-week lead times for larger orders. Pricing for U.S. delivery in 100,000+piece quantities starts at less than 4 cents each. For more information, visit the Vishay web site at www.vishay.com.

Taiyo Yuden (USA) Releases 0201 Case Size Capacitors for Cell Phone Modules and High Pin Count IC Applications

Taiyo Yuden (USA), Inc., announced the availability of its new RM Series of 0201 case size Multilayer Ceramic Capacitors (MLCC). The extremely small size of the RM Series capacitors results in higher component placement density, efficiency and PCB real estate savings, an especially critical consideration for designers of smaller, more functional portable communications and information processing devices. Typical applications include usage in cell phone modules and around integrated circuits (IC) of increasingly higher pin counts. The

company said its RM Series MLCCs provide the industry's highest capacitance value (i.e., Y5V 0.1μF).

At approximately one-third the footprint area of a 0402 capacitor—0.5mm² versus 0.18mm²—and nearly one-fifth the volume, Taiyo Yuden is one of only a few companies manufacturing such extremely small capacitors. The RM Series 0201 case size parts are currently available in production quantities for both Class 1 temperature compensating capacitors as well as Class 2 high capacitance value parts. Class 1 parts are available in E12 series from 0.5pF to 100pF. Class 2 X5R parts are available in E6 series from 100pF to 1,000pF; Class 2 Y5V parts are available in E3 Series from 22,000pF to 100,000pF.

RM Series MLCCs offer long component life, increased reliability and non-directional structure for simplified component mounting on the PCB. Compared to alternative technologies, such as Ta (not available in 0201 case size) and AE capacitors, the Class 2 RM Series capacitors employ base metal (Nickel) internal and external electrodes and have higher breakdown voltages. The RM Series' lower Equivalent Series Resistance (ESR) and Equivalent Series Inductance (ESL) minimize impedance at higher frequencies and reduce power loss and heat generation.

Price: From $0.015 each OEM quantities. For further information on RM Series 0201 case size MLCCs, contact Taiyo Yuden (USA) or visit the company's web site at www.t-yuden.com.

Cooper Electronic Technologies Expands PowerStor® Aerogel Supercapacitor Line

Cooper Electronic Technologies expanded its line of PowerStor® Aerogel Supercapacitors, to provide greater options and flexibility in pulse, main and hold-up power applications and memory back-up functions.

The aerogel supercapacitor has an extremely low ESR (equivalent series resistance) that allows it to be used where other types of super or ultra-capacitors have been unable to work.

PowerStor aerogel supercapacitors exhibit very low leakage current, allowing the devices to hold a charge for several weeks, if required. The aerogel supercapacitors have a long cycle life—virtually infinite compared to batteries—without any detrimental effects on performance.

The PowerStor A Series, featuring ultra-low ESR, offers capacitance values from 0.47 to 4.7 Farads and ESR values from 0.025 to 0.125 ohms. It is supplied in cylindrical packages and has a 2.5V working voltage. This device works well in low-duty cycle, high rate pulse power applications.

The PowerStor B Series Aerogel Capacitor is cylindrical and has a high specific capacitance, or high capacitance by volume, when compared to other brands. In a given size, it can store more energy with three times the capacitance of the original A Series by volume but only two times the ESR. B Series aerogel capacitors from 1F to 50F are available with ESR values from 0.025 to 0.125 ohms. Applications include main power, hybrid battery packs, hold-up power and pulse power in a variety of wireless communication devices, mobile computing equipment, industrial and power conversion systems.

The PowerStor F Series is an ultra-thin flat-pack design for space-constrained applications. It can be custom designed for mechanical fit and optimized for high capacitance and/or low ESR.

The PowerStor P Series, the newest addition to the line, features low ESR 5V aerogel supercapacitors with values ranging from 0.1F to 1.0F. This series combines the best features of high-energy rechargeable batteries and high-power electrolytic capacitors. Two types are available, the low ESR, high power PA Series with ESR values ranging from 0.15 to 0.2 ohms, and the higher energy density PB Series with ESR values ranging from 0.5 to 6 ohms. Low ESR allows high pulse current

Newsmakers

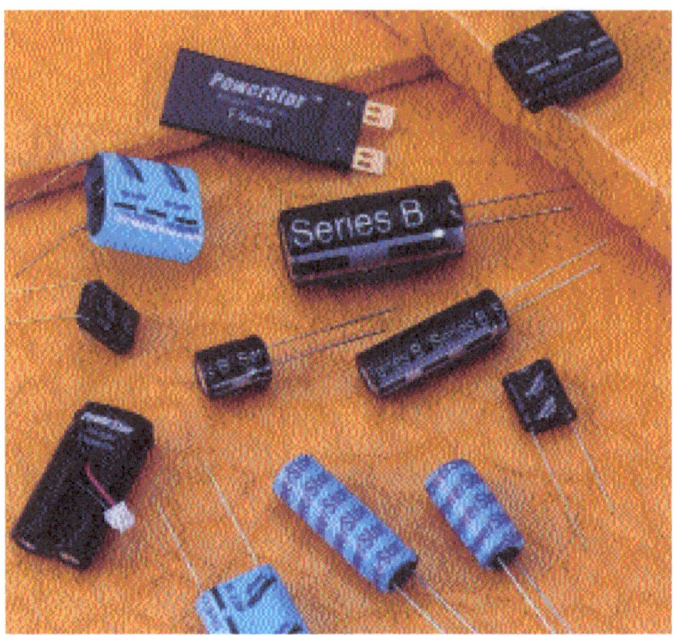

capability with minimum voltage drop. The P Series is ideal for portable devices requiring either hold-up or memory back-up or pulse power for a wireless transmitter, motor start or solenoid valve. Both the PA and PB Series are available with vertical or horizontal leads.

For more detailed product information visit the company web site at www.cooperET.com, or contact Marc Juzkow, product manager of PowerStor® products, at (925) 560-6742 or mjuzkow@cooperet.com.

Alpha-Core Introduces Smaller Toroidal Power Transformer Core

Alpha-Core, Inc., announced the "O"-Core, wound from grain-oriented silicon steel strip that is gradually pointed toward the ends, replacing the constant width strip used in traditional toroidal cores. The cross-section of the O-core is therefore circular like an O-ring.

The O-core transformers are 15% smaller in size and weight than traditional toroidal transformers. They are available up to 1200VA, 50-60Hz, or can be customized. O-cores are more difficult to produce and are typically priced 20% higher than traditional toroidal cores.

For more information, contact Ulrik Poulsen, (203) 335-6805, or sales@alphacore.com, or visit www.alphacore.com.

EPCOS announces PTC Thermistor Innovations

EPCOS, Inc., announced several additions to its thermistor line. They include:

Temperature-limit sensing compact package

The EPCOS PTC Thermistor SMD Sensor A601 in a compact 0603 package provides advanced miniaturization and even faster response than the 0805. Designed with lead-free tinned terminations enabling use in reflow soldering, these thermistors limit temperatures in applications such as DC/DC converters, power supplies, electronic ballasts, home appliances and automobiles.

These components deliver a significant signal when a certain temperature threshold is exceeded. In fact, the resistance more than doubles within 10°C, making it easy to evaluate the signal. Size 0603 PTC thermistors have the same electrical specifications as EPCOS PTC Sensors A701 of size 0805, so they can be used as a drop-in replacement. Standard delivery mode is blister tape packaging.

Heater series thermistors for 42V automotive applications

EPCOS has developed two series of PTC heaters optimized for the 42V power net, which is already incorporated in the design of future car models. These heaters consist of round disks with a 12mm diameter and rectangular PTCs sized 35mm x 6.2mm x 1.4mm. Silver metalization makes these components ideal for clamp contacting.

Because of their self-regulating properties, heating devices based on PTC thermistors can generally eliminate control and regulating components as well as excess temperature protection. The maximum surface temperature is limited by the R/T-curve, so there are no glowing parts and the circuitry is nonflammable. PTC heating elements are ideal for automotive applications where safety and durability are essential. Examples include additional cabin heating, nozzle heating, diesel fuel preheating and deicing (e.g. blow-by).

Upside PTC and Telephone Pair Protectors

EPCOS introduced two completely new designs for telecom line protection. As with most EPCOS telecom PTCs, these components are compliant with ITU-T K20, K21 and K45 for basic level lighting surges (10/700 µs), level power induction (600V, 1A, 0,2s) and power contact criteria (230V, 15min). EPCOS telecom PTCs also withstand connection to main voltages 110/230VAC in tripped (high ohmic) condition.

Square PTC R212 for vertical SMD placement

Designed with the latest generation of MDF (Main Distribution Frame) modules in mind, the new square PTC R212 for vertical SMD placement offers significant space-saving advantages compared to other solutions. Originally designed for soldering contacts, the electrode of the R212 is also suited for clamp contacting.

New telecom pair protector

The design for line cards is driven by miniaturization and simplification. The new telecom pair protector

Continued on page 36

EDS 2003
IN ELECTRONIC DISTRIBUTION, EDS IS YOUR BRIDGE OVER TROUBLED WATERS.

WHERE DECISION MAKERS MEET

ELECTRONIC DISTRIBUTION SHOW AND CONFERENCE
the channel event for components, instruments, datacom, MRO, production

May 12th – Keynote Address featuring Juergen Gromer, CEO, Tyco Electronics
May 13th–15th – Exhibits, Conferences, Workshops, Networking
Las Vegas Hilton Hotel – Las Vegas, NV USA

When times are tough, EDS becomes more important than ever. Whether you are a distributor, a manufacturer, or a manufacturers' representative, you can't afford to pass up the chance to see everybody who is important to the future of your business. You can't afford to miss out on the new information, new contacts, new opportunities, that EDS provides.

At EDS, you see more people in less time, and accomplish more, than any other way, any other place…the high-level people you don't normally get to see during the course of the year. Everybody who is anybody is in Las Vegas for EDS.

EDS is about building your business. More of a people event than a product event, EDS is for building relationships, building trust. In today's troubled world, building those bridges is more important than ever – so it's more important than ever to be at EDS 2003. Apply for your badge today.

ELECTRONIC DISTRIBUTION SHOW CORPORATION
222 SOUTH RIVERSIDE PLAZA, SUITE 2160 · CHICAGO IL 60606
voice: 312.648.1140 · fax: 312.648.4282 · www.edsc.org · www.edsconnect.com · email: eds@edsc.org

Continued from page 34

(TPP) houses two EPCOS telecom PTCs for line protection. Suitable for SMD placement, the new TPP saves up to 40% of space compared to existing solutions, resulting in significant potential cost savings. The interior PTCs resistance match, so the designer no longer need worry about using two PTCs with similar resistance to establish line balance.

For samples or pricing information on any of these devices, call EPCOS at 800-888-7728.

Gowanda Electronics Announces New Surface Mount Inductors for High Current Power Applications

Gowanda Electronics introduced a series of surface mount inductors designed for use in power applications where high current handling capability and surface mount packaging are required. The new SMP 5057 Series of surface mount inductors is specifically targeted for use in DC/DC converters and switching regulated power supplies. The rugged construction enables the SMP 5057 inductors to be used in these and many other power supply applications where higher current handling capability and durability are important.

Specific applications include microprocessor and distributed power applications–including power supplies, power regulators, and DC to DC converters–especially in test & measurement equipment, automotive electronics, medical diagnostic equipment, and industrial laboratory analysis equipment.

Technical specifications for products in Gowanda's SMP 5057 inductor series include: inductance from 3.9 +/- 20% microHenries (at 1 kHz) to 100,000 +/- 10% microHenries (at 100 kHz), current ratings from 9.75 to 0.090 DC Amps and saturation current from 24.0 to 0.146 DC Amps.

Gowanda Electronics also offers custom designs in the SMP 5057 series in order to meet the specific requirements of an application. For design details or custom requirements contact Gowanda Electronics at (716) 532-2234 or check the company web site at www.gowanda.com. Pricing for the SMP 5057 series is $1.60/unit (US Funds) in production quantities.

For more information, visit www.gowanda.com.

Sprague-Goodman Enhances SURFTRIM® Line of Surface Mount Trimmers

A new series of miniature, low profile surface mount trimmer capacitors has been added to the Sprague-Goodman SURFTRIM® line. Models in this new GKRP series measure only 3.2 x 2.5mm (0.125 x 0.098"), with a height of less than 1.3mm (0.051") above the mounting surface. A thin silicone coating for protection allows

usage for assembly processes that include washing.

Four capacitance ranges are offered: 3.0-5pF, 3.0-10pF, 5.0-20pF and 7.0-30pF. Q is 150 minimum at 1 MHz. Temperature coefficient of capacitance is NPO ±500 ppm/°C for the two lower ranges, and N750 ±500 for the two higher ranges. Operating temperature range is -40°C to +85°C. Voltage rating is 25 VDC, and dielectric withstanding voltage is 75 VDC. Insulation resistance is 10^4 Megohms minimum. Tuning torque is 10-100 g-cm (0.14-2.08 oz. in.).

Carrier-and-reel packaging is standard. The unit price is $0.49 each for 2000 pieces supplied on a 7" (18 cm) reel. Lead time for delivery is 12 weeks.

Continued on page 38

process control equipment. Other applications include use in security systems, instrumentation, bar code and

Better Prices, Superior Service:
It's All a Matter of Distribution.

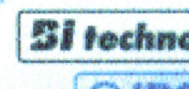

ECA's Distribution WhereHouse is a comprehensive online catalogue of products available through authorized distribution. The WhereHouse complements ECA's programs that promote the distribution channel as a major sales and marketing vehicle for manufacturers. Users can search by product or manufacturer.

ECA Resource Central is the one place where you can find any information you need about electronic components in real time. The Distribution WhereHouse makes the product distribution channel visible to customers. It also provides an opportunity for ECA members to market their products.

To find out more, visit the ECA Resource Central Distribution WhereHouse.

www.ec-central.org

2500 Wilson Blvd.
Arlington, VA 22201
Tel: 703-907-7070 · Fax: 703-875-8908

Newsmakers

Continued from page 36

For detailed technical information, contact Bernice Feller at Sprague-Goodman Electronics, Inc., (516) 334-8700, e-mail: info@spraguegoodman.com, web site: www.spraguegoodman.com.

Vishay Releases Thin-Film, Single- and Multilayer High-Density Interconnects

Vishay Intertechnology, Inc., announced the release of new thin-film, single- and multilayer high-density interconnects (HDIs), low-noise products with enhanced signal routing and response conditioning that integrate conductor patterns and other passive components in custom resistor solutions.

Each of these patterned, thin-film substrates is designed by Vishay Electro-Films (EFI) with the customer, the two working as a team to develop application-specific solutions for use in microwave circuitry and

hybrid circuitry in high-performance, low-noise power amplifiers; avionics; satellites; and medical instrumentation. Because of the high complexity of these devices and their close relationship to end-product performance, this team-oriented approach ensures the optimum ratio of price to performance.

Vishay EFI has had distinctive success in developing and manufacturing rugged and well-defined air bridges down to a 0.001in. width with consistent air-gap dimensions. Because multilayer HDIs are built up rather than out, they are capable of integrating a greater number of components into a much smaller footprint. Dimensions for each device may be as small as 0.02 in. by 0.02 in. or as large as 4 in. by 4 in., with thickness ranging from 0.005 in. to 0.050 in.

The new HDIs may be manufactured in designs of up to five layers and with special shapes, vias, and patterns. Each is available with a nichrome or tantalum nitride resistor element and in a wide variety of materials and conductor and adhesion metals. Additional options include metallized through holes, backside metallizing and patterning, wrap-around patterned edges, thick copper power line conductors, filled vias for added low-thermal-conductivity paths to a ground plate heat sink, and both aluminum and gold wire bond pads on the same substrate to provide monometallic interfaces in very-high-temperature applications. Multi-level metallization is achieved using polyimide insulation. Available substrates are alumina (Al_2O_3), beryllium oxide (BeO), aluminum nitride (AlN), quartz, and silicon.

Per piece pricing varies depending on quantity and design complexity. Readers may obtain a quotation by e-mailing drawings or requirements to sales@electro-films.com. Additional information about Vishay and its products can be found at www.vishay.com.

Littlefuse Develops SMD Fuse for Lead-Free Manufacturing Standards

To meet the growing needs of higher volume product manufacturers who must meet emerging worldwide lead-free manufacturing requirements, Littelfuse, Inc., introduced the 467 Series SlimLine™ Very Fast-Acting Fuse. The smaller EIA 0603 size targets applications requiring overcurrent protection in a fast response, smaller footprint package. The low profile fuse is constructed of high performance substrate and packaging materials designed to withstand the 260 degree ambient temperatures used in lead-free soldering operations. The 467 series complies with present electronic industry manufacturing standards for lead-free circuit board content including IEC 60068-2-20 as well as with

the EIA-IS-722 (Low Voltage Supplemental Fuse Qualification Specification), CSA and UL approvals.

As opposed to other planar fuses, this design uses no lead in either the fusing element or its termination,

Continued on page 40

The only thing subject to change is the weather.

Heraeus CT2000 LTCC technology performs in any environment.

Even if our intrepid explorer's next stop is the Amazon jungle, he'll get great reception. Because his mobile phone contains filters and modules made with Heraeus' CT2000 LTCC technology, with an incredibly low T_f of <10ppm/C.

T_f stands for temperature coefficient of resonant frequency, and it means that devices made with CT2000 LTCC materials perform consistently in the toughest environments, from over the Pole to under the hood.

As usual with Heraeus, there's more. Our new high-Q, pure silver conductors combine with CT2000 to support the most advanced applications—such as Bluetooth, 802.11, or other modules operating beyond 5GHz. And along with products, we supply expertise to help your process run smoothly.

So come in from the cold for a great reception: call or log onto Heraeus today.

Expect more from Heraeus.
CIRCUIT MATERIALS DIVISION
24 Union Hill Road, West Conshohocken, PA 19428
Tel: 610-825-6050 • Fax: 610-825-7061 • Visit us on the Web at: www.4cmd.com
CT2000 LTCC technology produced under license from Motorola.

Continued from page 38

which is composed of 100% copper/nickel/tin for lead-free compliance. This is the second major lead-free product release by Littelfuse, which has pledged to reduce or eliminate lead in all of its product lines. The company's goal is to allow end-product manufacturers to comply with emerging world-wide lead free initiatives such as Europe's Restriction of the use of Certain Hazardous Substances in Electronic and Electronic Equipment, (RoHS) and the Japanese Electronics Industry Development Association lead-free product roadmap.

Created for applications requiring extremely fast opening times to protect sensitive circuitry from momentary high levels of current, the 467 Series can cut off flow to a circuit in as little as .02 seconds at 300% of rated current, and in less time at higher surge current levels. The 467 series provides superior inrush withstand characteristics (I^2t) over ceramic or glass-based chip fuse products. The standard outline 0603 device measures only .305mm (.012″) in height and can be used in space-restricted environments on either AC or DC circuits. For users of previous Littelfuse 0603-sized fuses, mounting pad and electrical performance is identical to the Littelfuse 431 and 434 series fuse. The fusing element protective coating material is resistant to industry standard aqueous and solvent-based cleaning operations. The 467 series is available in tape-reel packaging. Pricing for all amperage ratings are $0.31 for 10,000 piece quantities.

Additional electrical parameters and performance characteristics are available at www.Littelfuse.com. Customers can obtain more information or receive samples by contacting a Littelfuse application engineer at (800) 999-9445.

Kamaya Widens Availability of Carbon Composition Resistors

Normally, a leading manufacture of SMT resistors would not promote carbon composition resistors as a leading technology. Yet, after more than 50 years of production, the demand for these trusted devices is increasing in certain applications that need their ability to dissipate extremely high energy pulses. Because of this, one of the remaining large-scale manufacturers of these devices, Kamaya Inc., has decided to re-emphasize the product line and expand its availability.

"Although we've got the newest resistive technologies on earth, like our new sub-miniature 0201 series, we are getting more and more inquiries for carbon composition," notes Mike Liebing, Kamaya's VP of sales and marketing in North America. "Customers like the idea that a QS9000 manufacturer like Kamaya is producing carbon composition resistors. "The bottom line for these manufacturers is that there's very little out there that has the pulse handling capability of carbon composition

at that price, particularly when compared to thinner element resistors." Liebing admits that many other resistor technologies provide a more accurate solution, but he says most of the manufacturers that depend on carbon composition accommodate its drift profile in their designs. He says larger users are in industries including telecom/ datacom, televisions, industrial motion-automation controls, medical, power supplies, circuit protection, appliances, HVAC, lighting, defense, and traditional TV design among others.

Liebing says no product advancements are planned, but that production capacity is being made available for increased market needs. Distributors for the component have been added. The company is planning a new promotional effort and the product is being featured in more detail on the company web site (www.kamaya.com) as well. Kamaya offers a full resistance range from 1 ohm to 22M ohm with power dissipation up to 1/2 watt and rated voltage as high as 350 volt. Tolerance ranges down to 5% are also available. Prices for the product are available through local distributors listed on the web site.

Frontier Electronics Introduces High Current, Low Profile SMD Power Inductors

Frontier Electronics introduced the Model CSH 1250S/CSH Series of Low Profile Power Inductors. Windings consist of a flat stock conductor, wound perpendicular to the axis of the core. This produces the same effect as a planar inductor without the space lost in PC board insulation.

The Model CSH 1250S/CSH Series dimensions are 12.0mm x 12.0mm high (CSH1250S) and 12.5mm x 12.5mm (CHS1260S), making them ideal for high current ultra-small DC-DC Converter applications such as encapsulation within the same package as other devices.

Specifications include: low profile, maximum 5.6mm high; magnetically shielded; operating temperature –30°C to 100°C; inductance range from 0.8µH (15.0A) to 7.0µH (5.6A) for the CSH1250S and 1.5µH (16.5A) to 10µH (7.6A) for the CSH1260S; DCR from 3mΩ to 16.8mΩ. Price is 70 cents each in large quantities.

For more information, contact Jeannie Gu at (805) 522-9998 or Jeannie@Frontierusa.com. ❑

LTCC Components and Modules

World Markets, Technologies & Opportunities: 2003-2007

The new, 170 page report from Paumanok Publications, Inc., that assesses global markets for low temperature co-fired ceramic components and modules.

This Report Includes:

- **Supply Chain Analysis: Raw Materials, Tapes, Metallizations.**
- **Value and Volume of Global Demand by Product Type.**
- **Average Unit Pricing and Price Trends.**
- **Competition From FR4.**
- **Demand by End-Use Market Segment.**
- **Competitive Environment by Product Type & End-Use Segment.**
- **Detailed Forecasts, Trends & Directions.**

Report #010301
LTCC Components & Modules
World Markets, Technologies & Opportunities: 2003-2007
Price: $2,400.00 USD • Pages: 160 • Release Date: January 2003

For a complete table of contents and brochure, please contact:

Paumanok Publications, Inc.
(919) 468-0384
(919) 468-0386 Fax
email: info@paumanokgroup.com

ADVANCED PRODUCTS.

Tyco™ Electronics has introduced a new Raychem Circuit Protection PolySwitch® TSM600-250 overcurrent protection device for DSL and other telecommunications network equipment applications.

Features include:
- SMT form factor
- Maximum interrupt fault voltage rating of 600V_{AC}
- Lightning surge capability of 1kV, 100A, 10/1000μsec
- Low resistance
- Hold current of 250mA
- Two resistance matched PPTC protection components per package
- Small footprint

Find out more at www.ttiinc.com/ap

ADVANCED PRODUCT KNOWLEDGE

Henry is big on teamwork. A TTI passives specialist, he is thoroughly grounded in his products and up to speed on new technology. This kind of knowledge helps turn the supplier-distributor-customer relationship into a championship team. And like we said, Henry is big on that. With Henry's help, Tyco Electronics and TTI do whatever it takes to help the customer with timely, flawless orders.

Place your next Raychem Circuit Protection order with TTI, and see how teamwork like this works for you.

PASSIVES   INTERCONNECT   ELECTROMECHANICAL

1-800-CALL-TTI   TELESERVICES   1-800-ASK-4-TTI

• • •   www.ttiinc.com   • • •

www.ingramcontent.com/pod-product-compliance
Lightning Source LLC
Chambersburg PA
CBHW051103180526
45172CB00002B/759